W9-BWP-292

ILLINOIS CENTRAL COLLEGE
GF895.P37 1992
STACKS
Tropical rainforests /
A12900 758889

TROPICAL RAINFORESTS

Withdrawn

Tropical Rainforests presents the most up-to-date and wide-ranging review of the problems and prospects of the world's most complex and abundant ecosystem. The book examines where and how fast rainforests are being cleared, drawing on examples from all major forest areas. The consequences of clearance are examined at local, regional and global scales.

The author achieves a balanced overview of the current state of the world's rainforests, discussing both the consequences of clearance (for ecology, environments and peoples) and the possible solutions (such as conservation and protection, reforestation, sustainable management, changing tropical timber trade and international investment programmes).

Well illustrated with maps, figures and photographs and with a comprehensive bibliography, *Tropical Rainforests* provides an essential introduction for students of Geography, Ecology and the environment, teachers, environmentalists, development practitioners and the general public.

Chris C. Park is Senior Lecturer in the Department of Geography at Lancaster University. He has published widely on environmental issues.

TROPICAL RAINFORESTS

Chris C. Park

I.C.C. LIBRARY

London and New York

85571

GF
845
.P37
1992

First published 1992
by Routledge
11 New Fetter Lane, London EC4P 4EE

Simultaneously published in the USA and Canada
by Routledge
29 West 35th Street, New York, NY 10001

Reprinted 1994

© 1992 Chris C. Park

Typeset in Baskerville by J&L Composition Ltd, Filey, North Yorkshire
Printed and bound in Great Britain by
Biddles Ltd, Guildford and King's Lynn

All rights reserved. No part of this book may be reprinted or reproduced or utilised in any form or by any electronic, mechanical, or other means, now known or hereafter invented, including photocopying and recording, or in any information storage and retrieval system, without permission in writing from the publishers.

British Library Cataloguing in Publication Data

A catalogue record for this book is available from the British Library.

Library of Congress Cataloging in Publication Data

A catalog record for this book is available from the Library of Congress.

ISBN 0–415–06238–1
0–415–06239–X Pbk

To my parents
Alex and Margaret Park,

for giving me so much
and asking so little in return.

CONTENTS

PLATES

FIGURES AND TABLES

FIGURES

TABLES

ACKNOWLEDGEMENTS

It often strikes me as grossly unfair that only the author's name appears on the cover of a book, because producing it really is a team effort . . . even if different members contribute different things, some of which might seem bigger or more important than others. As ever, I am surrounded by most gifted and thoughtful people, and my team have helped me in many ways beyond what they are aware of.

The silent sufferers while a book is being written are usually the author's immediate family, and mine is no exception. In Angela I am blessed with an understanding wife who accepts my long days at the word-processor with little hostility and I thank her for a warm and supportive home environment in which to live and work. Sam, our son, accepts my slavery to the keyboard with characteristic maturity while Elizabeth, our daughter, likes to sit on my knee and pretend to write!

The Geography Department at Lancaster University has long proved to be a stimulating and friendly academic base, and I thank my colleagues for simply letting me get on with things! Claire Jarvis drew most of the figures, and – as ever – is a joy to have around.

Routledge have been patient publishers, and I thank Tristan Palmer for his continued support and encouragement, and for not being disappointed when I missed my first deadline. I am grateful to Miles Litvinoff for meticulous attention to detail at the copy-editing stage. Jonathan Porritt and Thomas Lovejoy offered useful suggestions for clarifying and improving the text, for which I am grateful. I assume responsibility for all errors that remain.

I thank those mentioned by name above, and our supportive circle of close friends (particularly Robin and Jill, Steve and Audrey – *they* know who they are!), for making the writing of this book such a pleasure.

Chris Park
Lancaster

1

THE TROPICAL RAINFOREST: HISTORY AND ENVIRONMENT

This is the forest primeval

Henry Longfellow, *Evangeline* (1847)

1.1 INTRODUCTION

Tropical rainforests are the most complex ecosystems on earth. Rainforests (better known to many people as jungles) have been the dominant form of vegetation in the tropics for literally millions of years, and beneath their high canopy lives a diversity of species which is unrivalled anywhere else on earth.

1.1a Images and impressions

For more than a century travellers have recorded vivid descriptions of the rainforests, which emphasise abundance and grandeur. Charles Darwin kept a detailed log of his impressions of the forests around Rio de Janeiro, which he visited in April 1832 during the voyage of the *Beagle*. He wrote:

> After passing through some cultivated country, we entered a forest, which in the grandeur of all its parts could not be exceeded. . . . The trees were very lofty, and remarkable, compared with those of Europe, from the whiteness of their trunks. . . . The forest abounded with beautiful objects. . . . The greater number of trees, although so lofty, are not more than three or four feet in circumference. . . . It is easy to specify the individual objects of admiration in these grand scenes; but it is not possible to give an adequate idea of the higher feelings of wonder, astonishment, and devotion, which fill and elevate the mind.[1]

Since about 1980 interest in the rainforest has shifted full circle. From being seen as a threat or nuisance, it is now widely seen as under threat, with mounting concern for its future survival. The place to be tamed and conquered is now viewed as a place to preserve and protect.

This about-turn in attitude has been triggered by the real threat of destruction and clearance of the rainforest habitat world-wide. Although clearance *per se* is not new, the pace of deforestation is faster than ever before, and there are real fears that within one generation there will be hardly any natural rainforest left anywhere in the world.

There is already a fairly sizeable literature on the character and dynamics of tropical rainforests.[2] In this book we will examine why the rainforest is so important, why it is being cleared, what the consequences of this clearance are and what solutions are available. But first we need to define exactly what we mean by the term 'rainforest', examine where it is found and reflect on why it is regarded as the most important ecosystem on earth.

1.1b Classification

The rainforest is one of several types of forest found throughout the tropics, and each type has different characteristics.[3]

The closed forests account for about half of the total area of tropical forest (around 62 per cent of the natural tropical forest) and comprise two types of continuous tree cover (Table 1.1). Eleven-twelfths of the closed forests, by area, are tropical moist forests and the rest are deciduous and semi-deciduous forests of various types. About two-thirds of the moist forests are tropical rainforests, composed of evergreen broadleaved trees which flourish in the high temperature and humidity of the low latitudes. The tropical moist deciduous forests (or monsoon

Table 1.1 Distribution of tropical forest types

	Forest type	*Total area (million km^2)*		
(1)	Closed forests			12
	(a) Tropical moist forests		11	
	(i) Tropical rainforests	7.3		
	(ii) Tropical moist deciduous forests	3.6		
	(b) Deciduous and semi-deciduous forests		1	
(2)	Open woodland			7.34
(3)	Fallow forests			4.10
(4)	Tropical forest plantations			0.115
	(a) Industrial plantations		0.071	
	(b) Non-industrial plantations		0.044	
TOTAL				23.55

Source: World Resources Institute (1988).

forests) grow on the fringes of the tropical rainforests, and lose their leaves in the dry season.

Most of the remaining tropical forests are open woodland, including shrublands and types of savanna, pasture and grassland which are partly wooded.

Almost all (97 per cent) of the tropical forests which have been modified by human activity are fallow forests, areas which have recently been farmed and then abandoned or left to regenerate naturally. Only a very small area is covered by tropical forest plantations. The industrial plantations produce commercial timber, pulpwood or charcoal; the non-industrial plantations are mainly for fuelwood production or environmental protection.

1.1c Distribution

The tropical rainforests provide a discontinuous belt of green around the globe, between the tropic of Cancer (23.5° north) and the tropic of Capricorn (23.5° south). Dense rainforest is the natural climax vegetation of the hot, humid tropical zone and it flourishes particularly in the lower latitudes (between 10° north and south of the equator). Just under half of the tropical zone (49 per cent according to the World Resources Institute)[4] is covered by forests (Figure 1.1).

Most of the tropical countries with surviving rainforests are developing countries, for whom the forests provide a valuable capital asset.

The total area presently covered by tropical rainforests is estimated at 12 million km^2, which accounts for nearly a third of the world's forests (covering roughly 30 million km^2).[5] The distribution of forests within the tropics is uneven, reflecting the distribution of land and sea and the impacts of this on climatic boundaries. The latitudinal boundaries of the rainforest are determined mainly by precipitation, while altitudinal limits are determined more by temperature. Some rainforests thrive beyond the 10° north and south latitudes, where high rainfall encourages forest growth. Such patches occur in Central America, the north-east coast of Australia and the great valleys of southern China.

The main rainforests today are found in three areas (Figure 1.1 and Table 1.2) – Latin America, Western Equatorial Africa and South-East Asia. Latin America houses the American Formation which is dominated by the Amazon and Orinoco Basins. It has over half (56 per cent) of the world total, much of it (3.31 million km^2, 48 per cent of the area's total) in Brazil and the rest in Peru, Ecuador, Colombia, Venezuela and French Guiana. Amazonia is the world's largest and most important surviving rainforest.[6]

The remaining rainforests are scattered in sixteen countries in West and Central Africa (18 per cent of the world total) and South-East Asia

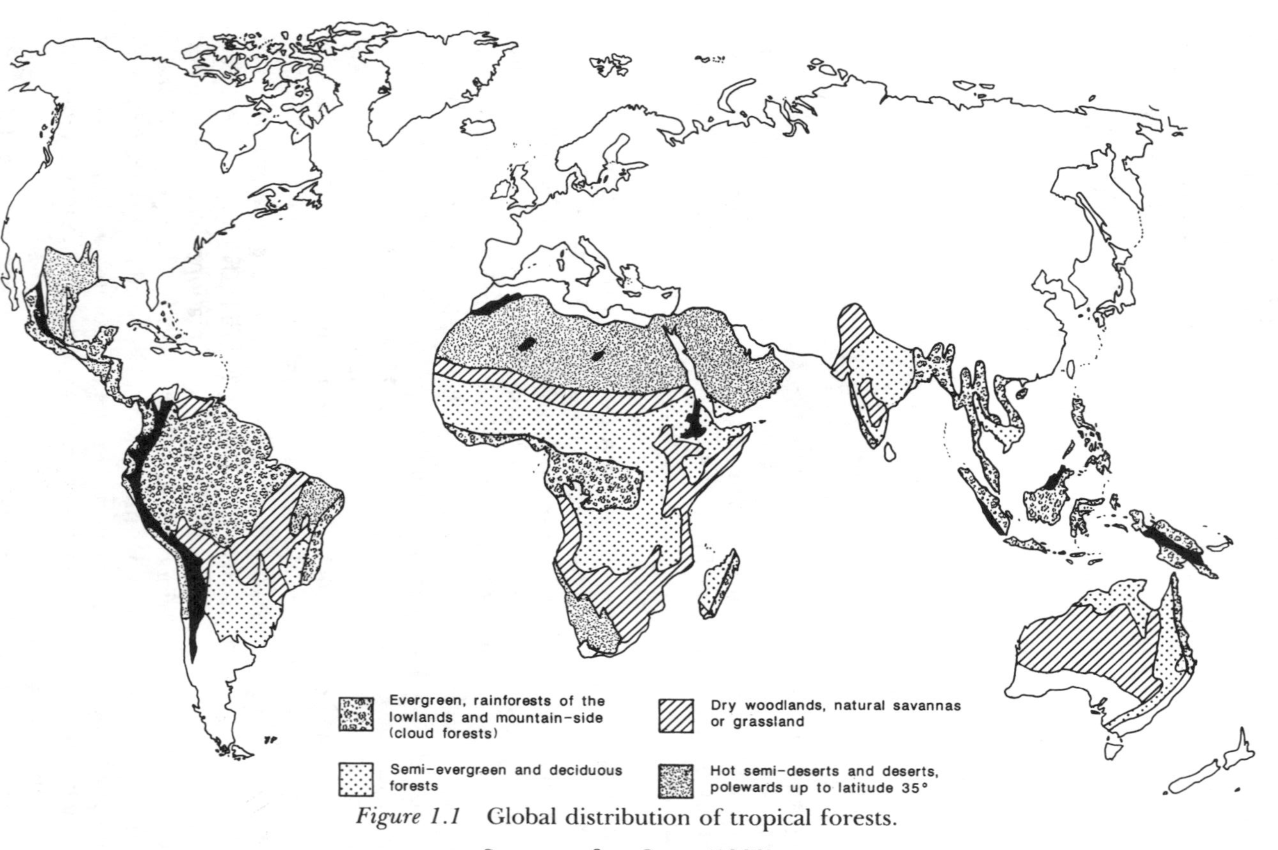

Figure 1.1 Global distribution of tropical forests.

Source: after Cross (1990)

Table 1.2 Distribution of tropical forests

Region	Land area	Closed forest	Open forest	Fallow forest	Total forest area
	Areas, in millions of km², excluding plantations				
Africa	21.90	2.17	4.86	1.66	8.69
Asia and Pacific	9.45	3.06	0.31	0.73	4.10
Latin America	16.80	6.79	2.17	1.70	10.67
TOTAL	48.15	12.02	7.34	4.10	23.46

Note: the total figures include rounding differences.
Source: World Resources Institute (1988), table 5.1.

(25 per cent of the world total). The African Formation includes the Cameroons and the Congo Basin in countries such as Gabon, Zaïre and Madagascar. The Indo-Malaysian Formation in South-East Asia includes parts of western and southern India, the Far East (especially in Indonesia – particularly Borneo and Papua New Guinea – which now has about 10 per cent of the world's remaining tropical rainforest) and north Australia.

1.1d Lack of reliable information

It is surprisingly difficult to define exactly what the total area of rainforest is today, for various reasons.[7] Some countries have better (and more complete) survey coverage than others, and not all of the figures available refer to the same year. Recent developments in remote sensing technologies promise much better survey coverage in the future,[8] although between 1970 and 1988 less than 60 per cent of tropical forests had been surveyed using any form of remote sensing, particularly available satellite technologies which are costly.[9]

Figures reported by the United Nations Food and Agriculture Organisation and by individual countries are often underestimates. Figures quoted by campaigning groups must also be treated with some caution, because such groups may well overestimate rates of forest clearance and underestimate the size of remaining areas. Estimates also vary because they include different types of forest cover. Some, for example, include coniferous and bamboo forests as well as broadleaved forests, whereas others do not. It is also difficult to distinguish between primary (original) and secondary (regrowth) forest cover, even in satellite images, so that quoted figures might not always be comparing like with like.

The most recent reliable information on areas of tropical forests comes from the World Resources Institute.[10] These figures (Table 1.2) provide a valuable framework for examining the distribution of tropical forests today and a useful baseline for assessing rates and patterns of deforestation (in Chapter 2).

The proportion of the earth's land surface covered by rainforest is small (around 8 per cent),[11] but its significance is considerable. Possibly half (up to 90 per cent according to some experts) of the total number of species of plants and animals found anywhere on earth are found either exclusively or mainly within tropical forests.[12] We will evaluate the wider significance of the rainforests in Chapter 4, particularly in terms of what would be lost if all remaining forests were to be cleared.

1.2 CLIMATE AND RAINFORESTS

Climate exerts a strong influence over the broad distribution of rainforests within the tropics. It also has a marked effect on regional patterns and structures of rainforest vegetation and habitat.

Rainforests occur in hot moist climates. They receive more solar radiation throughout the year than any other vegetation zone on earth, which promotes the rich variety and luxurious character of vegetation.

1.2a Characteristic climate

The typical rainforest climate has two main distinguishing features – relatively constant temperatures, and heavy rainfall. Tropical forests grow under a narrow range of temperatures but a fairly wide range of precipitation (Figure 1.2). The combination of these two climatic controls creates the very special environment for rainforest growth.

Temperatures in rainforest areas are high, and they vary relatively little throughout the year. Temperatures remain fairly constant at between 20°C and 28°C through the year,[13] the warmest months perhaps a degree or so higher than the coldest months in a given place. This uniformity occurs because the sun is mostly overhead, so variations in the length of daylight throughout the year are limited.

Temperature variations are often greater from day to night than from month to month, so that rainforest dynamics are more variable over short-term diurnal cycles than over longer seasonal or annual timescales. Diurnal temperature variations might be as high as 17°C. Intense heat during the afternoon (when temperatures may reach 35°C) gives way to cold at night (temperatures may fall as low as 18°C), then fresh conditions in the morning, in a never-ending rhythm of daily change. Many tropical areas have a fairly predictable daily cycle of weather dictated by temperature and humidity. Early morning mists (caused by cooling at night) evaporate as the sun rises, and by late morning convection currents start to rise from the forest. Clouds form and by

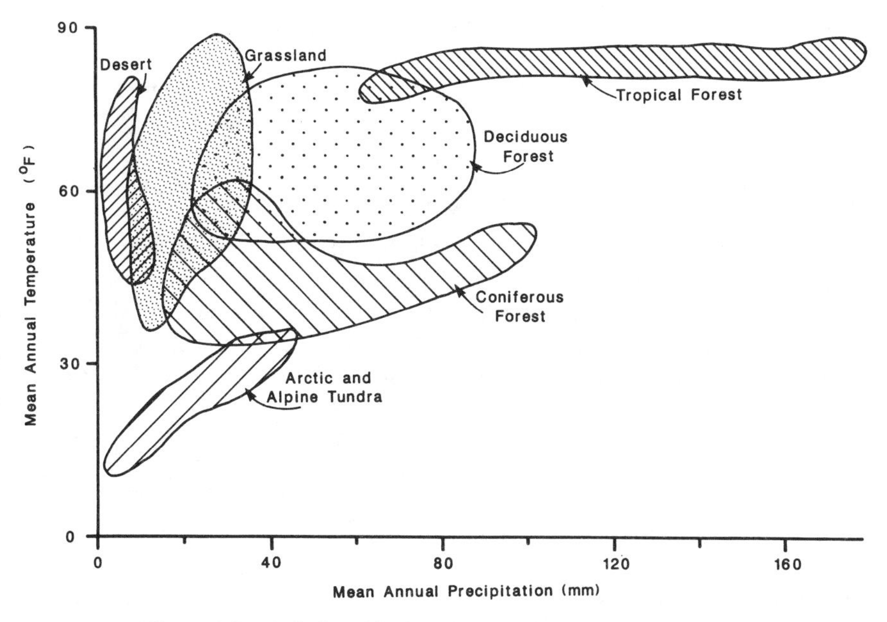

Figure 1.2 Relationship between vegetation and climate.

Source: after Cross (1990)

late afternoon these have often turned to storm clouds (with strong vertical thermal uplift), causing heavy rain often accompanied by thunder.

Heavy rainfall creates the specific conditions necessary for rainforest growth in the tropics. Indeed forest specialist Norman Myers[14] defines tropical forests as those 'forests with a mean annual rainfall of 1500 mm or more, and a mean monthly minimum of 100 mm'. Many receive between 1,750 mm and 2,500 mm of rainfall per year (rainfall in some areas can be as high as 10 metres a year).[15] Monthly rainfall in most rainforests is over 200 mm. Spatial variations in rainfall reflect topography and latitude; rainfall is highest over islands and coastal areas. It can also be high in mountain areas (such as western Amazonia).

The high rainfall arises mainly from thermal uplift. The Inter-Tropical Convergence Zone (ITCZ) lies over the rainforest areas for much of the year, causing warm moist unstable oceanic air to converge and rise. Orographic uplift gives rise to high rainfall in mountain areas.

Rain falls often in strong showers, and strong convectional conditions can bring heavy downpours, with intensities of 10 mm to 20 mm rainfall per day not uncommon (compared with 4 mm to 6 mm in London), and relatively little variation in intensity from month to month. There are a large number of rain-days a year in typical rainforest areas (for example,

Belem in Brazil has rain on up to 243 days a year), and thunderstorms are fairly common.

The high humidity in the tropics creates much cloud cover, averaging 50–60 per cent on most days. Cloudless days are rare.

1.2b Seasonality

Without doubt the most characteristic feature of the rainforest climate is the lack of pronounced seasonality. Unlike temperate areas, where the repeated rhythm of seasonal cycles creates a natural regularity of biological activity throughout the year, most rainforests have only one season (marked by high temperatures and rainfall) which continues incessantly. There is no summer and winter as such, only wet and dry seasons. Conditions are often uniformly hot and wet, with regular intense tropical downpours, although seasonal (monsoon) rainfall occurs in some areas.

This uniformity of climatic conditions means that rainforest plants grow, flower and shed leaves all year long, and that rainforest animals continue to reproduce and remain active throughout the year. There is no quiet time or rest period in the rainforest.

1.2c Zonal pattern

The tropical rainforest belt represents the hot, wet latitudes within a broad zonal pattern of world vegetation which is strongly influenced by climate (Figures 1.2 and 1.3). It gives way to tropical grassland (savanna), to north and south, in which the type of grass and type and density of trees vary according to rainfall. Desert areas (with rainfall below 250 mm a year) experience extreme drought and have little vegetation cover.

Beyond that, in the northern hemisphere, comes the Mediterranean vegetation (Figure 1.3), now mainly scrub but formerly forest (which has been cleared by burning and grazing). At higher latitudes still are the temperate grasslands (steppes, or prairies; now used mainly for grazing and wheat-growing) and deciduous forests (most of which have been cleared for cultivation), then the coniferous forests (taiga) and arctic tundra in which few hardy species can survive the extreme cold.

1.3 AGE AND STABILITY

Rainforests are without doubt among the earth's oldest and most stable ecosystems. These natural ancient monuments have outlived many early types of vegetation which we only know about from fossil evidence. Moreover, rainforest trees are widely regarded as living fossils themselves, exhibiting forms believed to be characteristic of primeval trees

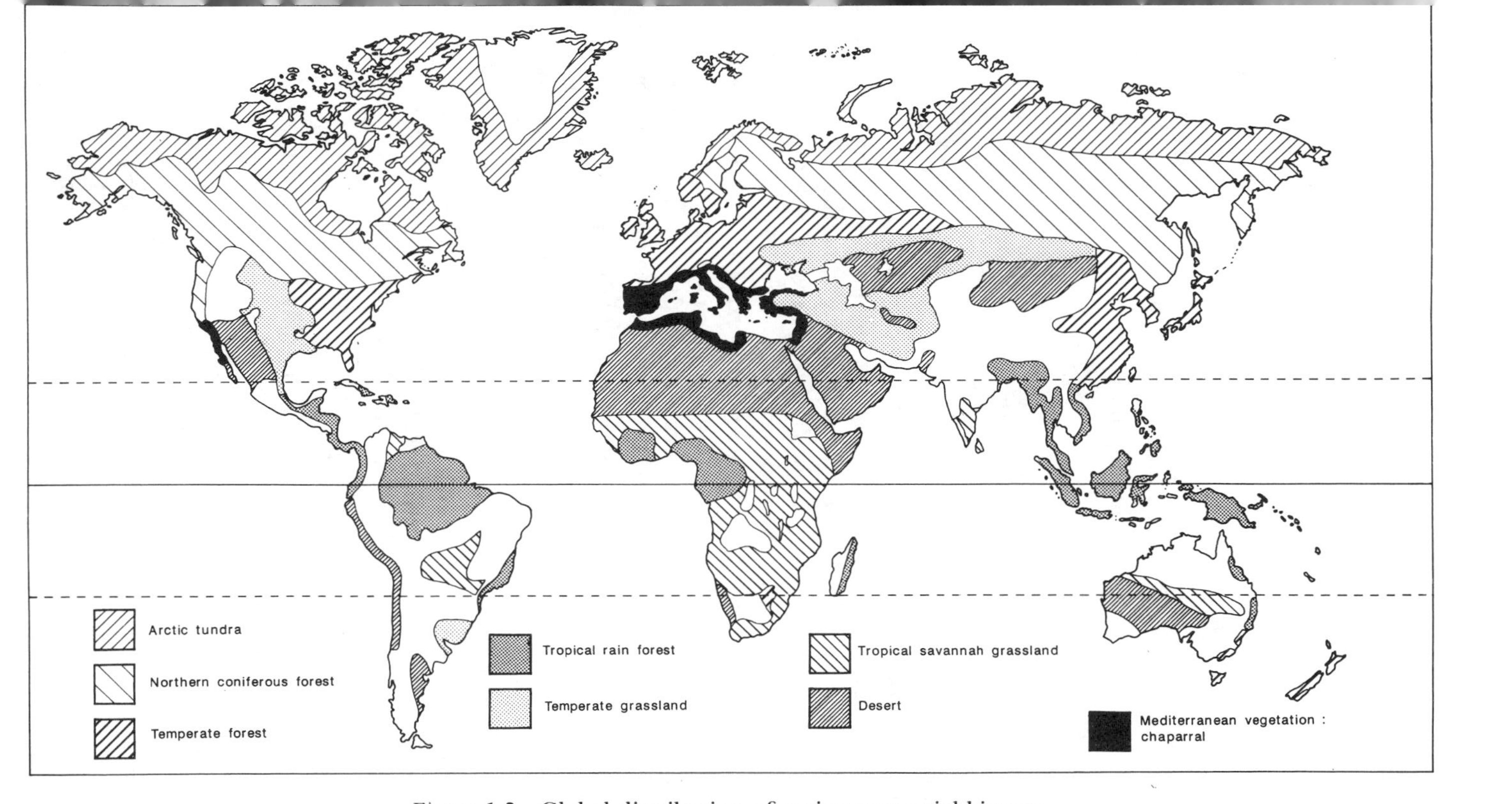

Figure 1.3 Global distribution of major terrestrial biomes.

Source: after Cox *et al.* (1976)

(thin bark, buttressed roots, tall light-seeking trunks). The rainforests certainly survived the Quaternary ice ages which destroyed vast areas of temperate forests, and provided critical ecological refuges in which species which escaped from the freezing wastelands further north could shelter.[16]

The distribution of rainforests today is largely historic, but doubtless more restricted than in the past.[17] Climatic changes over the last two million years have shrunk the area occupied by rainforests. Human activity, particularly since 1945 (see Chapter 2), has dramatically accelerated the shrinkage at the margins and is also devastating the interiors of many rainforest remnants.

1.3a Age

Quite how long the rainforests have existed in their current state and positions is widely debated. The estimates vary, but it is certainly a question of how many *millions* of years. Fossil and pollen evidence has revealed fragments of leaves and pollen grains similar to those in modern plants from as far back as 70 million years, and there is evidence that rainforest in parts of Indonesia has existed in more or less its present form for at least 60 million years.[18]

It is argued that some form of rainforests existed as far back as 150 million years ago, and that the Amazon appears to have remained relatively unchanged and undisturbed for over 100 million years. Fossil evidence suggests that exotic tropical forests covered much of the earth's land surface about 45 million years ago. Longfellow's description of 'the forest primeval' (in the opening quote) turns out to be remarkably apt.

There is widespread agreement that the main wet tropical regions of the earth have been stable for at least the last 40 million years. It is fair to assume, therefore, that rainforest development has been taking place since at least that date. Such long periods of relatively uninterrupted growth (during which other environments elsewhere have been repeatedly disrupted and disturbed) have allowed rainforest plants and animals to evolve and adapt, which in turn has allowed new species to appear. This is one explanation of the unusual richness and variety of rainforest species, each one naturally exploiting the myriad of ecological niches within the complex rainforest ecosystem.

1.3b Stability

The inferred longevity of rainforests is deceptive, because it suggests that they can survive against all odds. This is simply not the case. These ancient forests are in fact highly vulnerable to change, especially from external pressures. Dramatic and irreversible changes can follow from what appear to be relatively minor triggers (in the geological time-scale of events). Once the stability of the system is undermined or threatened, wholesale responses can follow very quickly.

What has taken tens of millions of years to evolve can be wiped out by human activities within a matter of a few human generations. Once it is gone, it is lost for ever. The natural rainforest ecosystems – vast, complex, diverse, productive – cannot be recreated by human design. Their loss represents an unprecedented waste of an irreplaceable biological resource. But the wastage is much wider than simply the loss of the trees themselves, because rainforests serve some critical environmental functions (see Chapter 4) which are fundamental to the future stability of the whole earth and its climate.

1.3c Uniqueness

The tropical rainforests are unique communities of plants and animals. They can appear, at first glance, somewhat random and chaotic, impenetrable tangles of vegetation characterised by intense and interminable competition. But close examination shows them to be highly regulated and tightly integrated ecosystems, in which everything (from the tallest tree to the smallest insect) has its rightful place, and everything depends on everything else. This is not to say that the ecosystems are simple; there are many complex relationships within them.

The rainforest ecosystem is important and irreplaceable. It is the most abundant source of life on earth, with an enormous species diversity, distinctive patterns of vegetation, a highly specialised system of nutrient cycling and rapid rates of growth. This complex ecosystem is finely adapted to a unique set of climatic and other environmental conditions.

The rainforests are reluctant to give up all their secrets. It has been noted that the rainforests are 'by far the richest, most diverse, and most complex biome on the planet, [but] they are also the least understood by science',[19] precisely because of this incomparable richness, diversity and complexity.

One particular hallmark of the tropical rainforests is their great density of plant material (biomass). An average 1 km^2 of rainforest might contain a similar weight of wood to 200–300 km^2 of temperate woodland, because rainforest trees are tall, wide and tightly packed. Rainforests have the greatest biomass (weight of living material per unit area) of all types of vegetation on earth (although biomass is highly variable and there are few data to generalise from), and they account for roughly half of the world's total biomass (estimated to be around 900 million tonnes).[20]

The rainforests grow under hot-house conditions, with high temperatures, limited seasonal variations in temperature, high humidity and high rainfall. Little wonder, therefore, that they are the most productive type of vegetation in the world. Mean net primary productivity in tropical rainforests is around 2,500 tonnes per km^2 per year, compared with around 700 tonnes per km^2 per year for boreal forests.[21] This productivity is double that typical of savanna grassland and twelve times that of deserts.[22]

1.4 APPEARANCE

The tropical forest has a very distinctive appearance and structure, dominated by its vegetation.[23]

1.4a Rainforest trees

Without doubt its most distinctive feature is the towering trees which soar skyward in search of light. Average tree heights may be around 50 metres; trees as high as 90 metres are not uncommon. In many rainforests the canopy is often 30 to 35 metres high.

The trees often look very similar, although there may be many different species, sizes and ages present. They are deciduous but evergreen in habit, continuously shedding leaves and growing new ones so their foliage appears permanent. This leaf-fall plays a critical role in nutrient cycling within the rainforests, and contributes to the creation of distinctive forest soils.

Species are highly adapted to cope with conditions found in the forest. They have quite distinctive large leaves, which are dark green to avoid excess moisture loss in the high temperatures. Many leaves have a thick cuticle with a wax surface, to allow the heavy weight of water to run off easily. Leaves on the lower trees often have leathery surface and accentuated downward-dipping points (drip-tips); as a consequence water reaches the soil below rather than evaporating off the leaves (Figure 1.4a).

The trees tend to have thin and straight trunks, with thin greenish-white bark and few low branches. Normally the only ones with branches close to the ground are short, young trees which are destined to grow much taller. Most of the branches grow towards the tree-tops, where they can obtain the maximum amount of sunlight. This gives rise to a very dense and compact vegetation structure, in which the trees grow close together. There may be between 400 and 600 trees broader than 10 cm diameter within a 1 ha patch of rainforest.[24]

Because the rainforest trees grow close together, their crowns form a leafy canopy which is usually high and closed (thus more or less light-proof) (Figure 1.4c). The canopy is formed by the dominant species within the rainforest, such as ebony, mahogany or teak. The height of

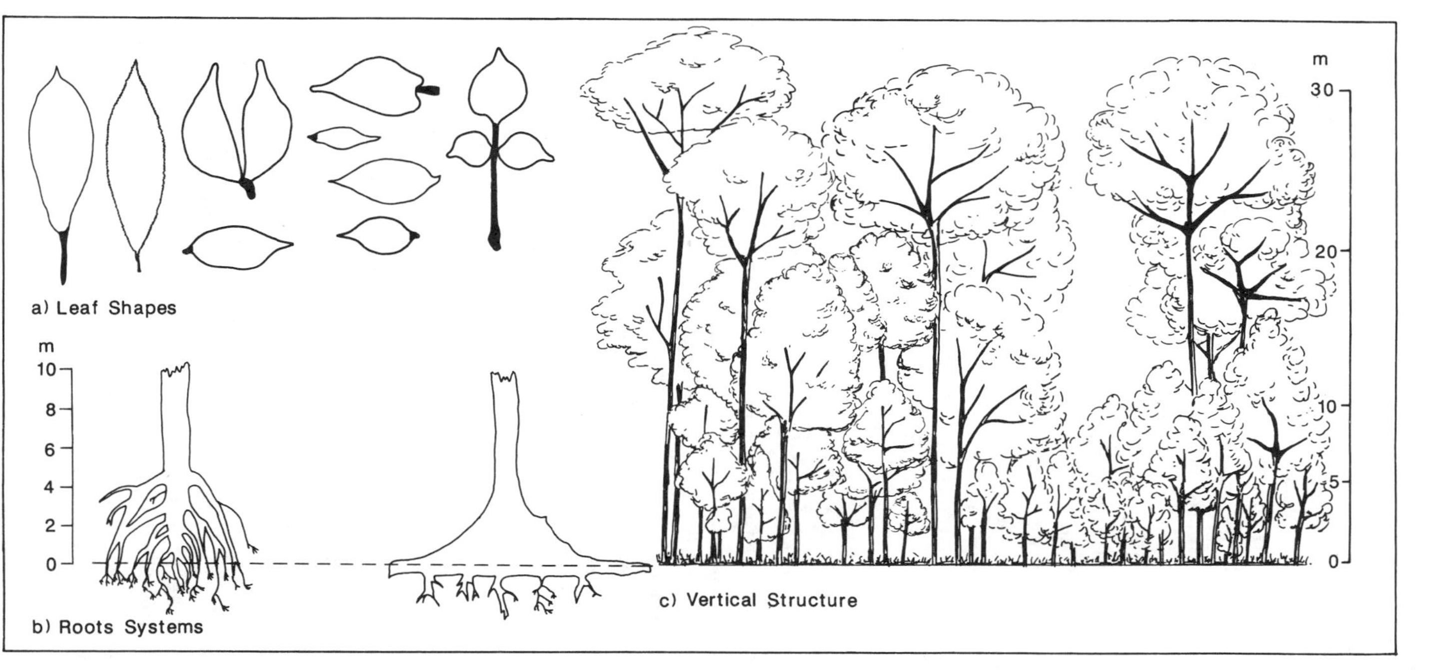

Figure 1.4 Vegetation characteristics in the rainforest, showing (a) typical leaf shapes with drip-tips, (b) stilt roots of *Uapaca* species (left) and heavy buttresses of *Piptadeniastrum africanum* (right) and (c) profile diagram of primary mixed forest. Moraballi Creek, Guyana (only trees higher than 4.9 m are shown).

Source: (a) and (b) after Goudie (1984), (c) after Pears (1977)

the canopy can vary greatly, but it is often more or less continuous over large areas (sometimes hundreds of km^2), giving the impression from above of effectively roofing the forest. Tall emergent trees pierce the canopy from place to place. The complexity of this canopy structure contributes to the great species diversity within rainforests, because it allows a large range of species of plants to coexist and provides a variety of micro-habitats in which all kinds of animals can live.[25]

A further hallmark of tropical rainforests is the great variety of tree species present, with no single species dominant within an area. Even a small patch of forest might contain literally hundreds of different species, unlike temperate forests which are very monotonous. Consequently the individuals of any particular tree species are often spaced widely apart throughout the forest, unless site factors (such as soil type, drainage, nutrient availability, aspect or micro-climate) make conditions particularly favourable for a limited number of species.

1.4b Climbing plants

Tropical rainforests also differ from temperate forests in their abundance of climbing plants, lichens, ferns and orchids. Many forests are literally festooned with hanging, climbing and creeping plants.

Particularly characteristic are the lianas (woody climbing plants) which twist like cables around the straight trunks of the tall trees. Lianas can be up to 200 metres or more in length, and they can entwine adjacent trees together like long, tough ribbons. They germinate in open ground after forest felling, and can quickly overgrow abandoned clearings to start the slow cycle of secondary growth.

The lianas play a key role in the rainforest, by effectively weaving the trees together into an interdependent mass. But this is something of a mixed blessing. They form links from tree to tree which can help to support individual trees, which generally have very shallow root systems and thus little anchorage. But if one or more trees fall naturally through wind-blow, or are cut down deliberately, they can often pull many others down or at least destabilise them.

Rainforest trees also provide a home for epiphytes like mosses, lichens and ferns. These are plants which grow on other plants (such as trunks and branches of trees) rather than in the soil (Figure 1.5), and they get moisture and nutrients from the humid atmosphere in the enclosed canopy environment. They are not parasitic on their host trees, and cause them no harm. Few epiphytes can tolerate the shaded conditions at ground level, and – being light-demanding – most thrive under the sunny conditions common in the tree canopy.

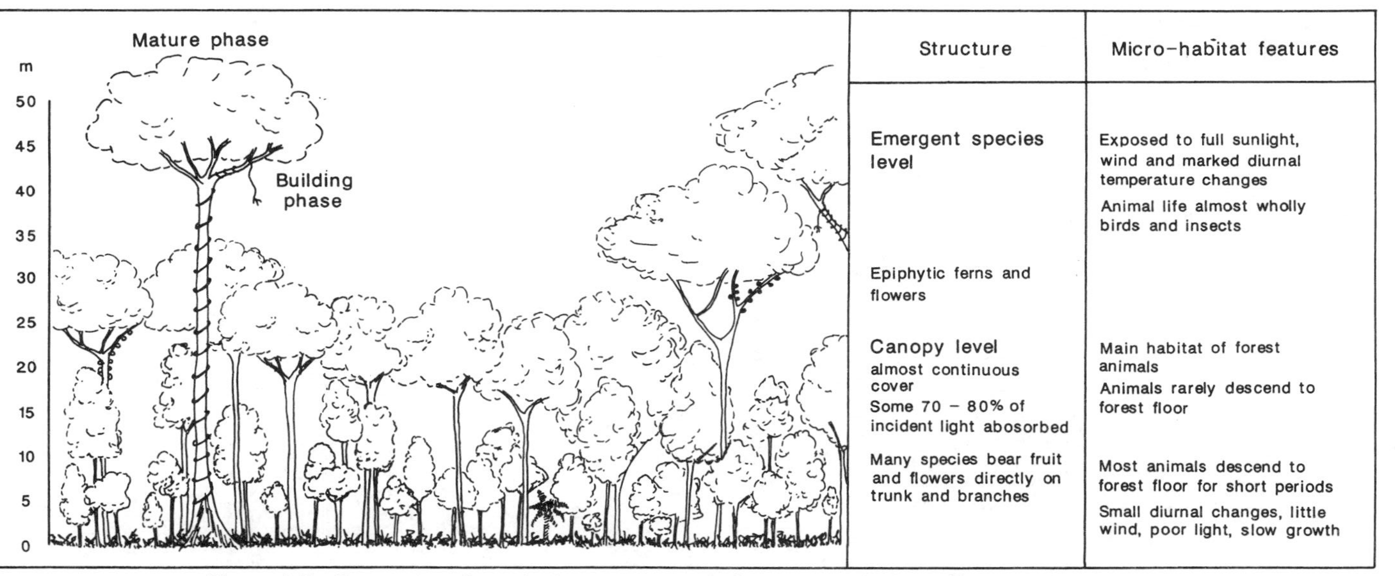

Figure 1.5 Structure of tropical evergreen rainforest (a typical profile).

Source: after Collinson (1977)

1.4c Roots and buttresses

Rainforest trees also have very characteristic root structures. Root development is generally very limited, so even the largest trees have neither deep nor large roots. For example, in the Ivory Coast rainforest 95 per cent of the trees' root systems are found in the top 130 cm of soil.[26] The huge weight of the tall trees is supported much more by neighbouring vegetation (effectively chained together by climbing lianas) than by the plant's own roots. As a result rainforest trees are highly susceptible to uprooting.

Many rainforest trees put out roots above the ground, and others have flying buttresses (Figure 1.4b). These shallow root systems appear to be a natural adaptation to the shallow soils common in tropical forest areas. Many trees have between three and ten buttresses radiating out from the lower trunk, perhaps a metre high and several metres long, giving the impression of a dense tangle of tentacles.

The precise function of these buttresses is not clear. They may well serve to increase resistance to wind-blow, but they probably also help the trees to gather moisture and nutrients. The curious root systems appear to be well adapted to absorbing most of the throughfall moisture.

1.4d Sunlight

Conditions within the forest are heavily influenced by the availability of light. Darkness normally prevails below the tree canopy, and all types of vegetation (including trees) compete for a share of the available light. This encourages the trees to grow tall and straight and to have wide umbrella-like canopies in order to exploit available light conditions to the full (Figure 1.5).

Most of the sunlight is very effectively filtered by the forest canopy. Less than 10 per cent (some studies put the figure as low as 1 per cent) of the sunlight which reaches the canopy normally reaches the ground below, and undergrowth is normally limited in such dim conditions. The ground beneath the rainforest canopy usually has very sparse cover.[27]

Bright shafts of light pierce through the forest roof from place to place, simulating some grand natural cathedral and stimulating the growth of isolated patches of herbaceous plants. Thick undergrowth and high densities of small trees only normally grow as the early stage of forest regeneration (as secondary forest) after an area has been cut down or fallen naturally through wind-blow.

1.4e Continuity and competition

The evergreen rainforest trees are rarely leafless because the lack of seasonality in the climate allows them to flower, shed old leaves and grow new ones at the same time. As a result the appearance of the forest changes little through time, and the light-proof forest canopy persists throughout the year. There is no quiet time in the rainforest.

The rainforest is a highly dynamic community in which plants die and new ones grow up to take their place. Its never-ending cycles of growth and decay (leaf fall, germination, growth, flowering, fruiting and dying) promote unrivalled diversity and productivity. They also create a complex web of interactions between individuals and species, competing and co-habiting in the rainforest hot-house.

The struggle for survival within the forest is relentless, with all plants competing for available light, air, space and nutrients. The variety of habitats within the forest vegetation supports a great diversity of insect and animal life, much of which is highly specialised, with life-cycles linked to particular plants. Competition between species within the favourable forest micro-climate means that few species dominate and many proliferate in close contact with one another. Diversity may also reflect a varied history of disturbance by human activity, particularly in the Amazon.

1.4f Vegetation stratification

The vegetation in the rainforest is highly structured, both spatially and vertically. The patterns are influenced by a range of factors, the most important of which is the availability of sunlight. Most types of forest show signs of stratification, in which vegetation is structured into vertical layers which maximise the efficiency of forest processes.

The traditional view[28] is that five layers can be recognised in the vertical structure of the typical rainforest, the upper three representing trees of differing sizes, ages and appearances, and the lower two representing the undergrowth (Figure 1.5).

A typical sequence, from the tree-tops to the forest floor, would have the following characteristics:

(5) The upper tree layer: the canopy of the tallest trees, usually higher than 25 metres. The trees usually have wide, umbrella-shaped crowns which are fully exposed to the sunlight and often form a continuous canopy. These dominant trees have smooth branchless trunks supported by large plank-buttress roots. Gaps between the trees allow sunlight to penetrate through to layers below. Isolated trees can grow much higher than the norm (sometimes as high as 70 metres) and these emergents tower above the surrounding canopy level.

(4) The middle tree layer: a sub-canopy of more tightly packed trees perhaps 10 to 25 metres high. They have narrower crowns because they are seeking light in the gaps between the higher trees. The canopy here is often more continuous than the one above because the tree-tops are

closer together, although a small amount of light still filters through this forest roof to the layers below. These medium-sized trees are often bound together by epiphytes and climbers. Many are young individuals of the dominant species, which are still growing up towards the canopy; others are fully grown trees adapted to darker conditions of permanent shade. Tree ferns and lianas complement the trees in this layer.

(3) The lower tree layer: lower trees, commonly between 5 and 10 metres high, very closely spaced. These usually have long, tapering crowns, which effectively plug any gaps in the canopy above and thus more or less completely shade out everything below. Most trees here are also young individuals, which in time will grow to the middle and upper layers.

(2) The shrub layer beneath the tree canopy: the lowest 5 metres or so, normally populated by isolated short, young trees (saplings) as well as shrubs and seedlings. Lack of light and room gives rise to small, thinly developed and widely spaced vegetation. Ferns and large herbaceous plants grow in the small spaces between trees in the higher layers.

(1) The herb layer on the forest floor, which is usually more or less bare, except for a few small plants, mosses, ferns and saplings on the ground, and a thin layer of leaf litter which has fallen down from above. Conditions are much more uniform here than higher above, because near the ground there is less air movement and so temperatures are lower and humidity is higher.

Each layer is slightly different in density, temperature, humidity and light levels, so each storey has its own characteristic group of insects and animals.[29] Some forest species exploit more than one layer. For instance, certain plants (the climbers and epiphytes) are found in all layers, and they often grow up and down between them exploiting variations in local conditions.

Botanists stress that it is misleading to think of a series of distinct layers within this vertical structure, as if the rainforest conforms to some immutable design. The layers are usually defined on the basis of statistical analysis of tree heights and frequencies, and the fivefold stratification is simply a convenient way of summarising the structure which recognises the basic ecological processes at work within the forest. Average heights of the different layers also vary from one forest to another, so the figures quoted are illustrative rather than definitive.

1.5 ANIMALS AND FOOD WEBS

1.5a Rainforest animals

Animals and birds are obviously much more mobile than plants, and some of them exploit conditions (including food supplies) in different layers within the forest (Figure 1.5). But relatively few animals move between the floor and canopy; most are adapted to live in and get food from specific parts of the forest. The species present naturally vary from place to place, but the structure and dynamics of different rainforest ecosystems are similar.

The trees themselves provide a home for some animals. Canopy-dwellers include many types of monkey, sloths, squirrels, rats and mice. Some, such as the flying squirrels, rarely descend to the ground. Others, like tree squirrels, move freely up and down the tree trunks between the canopy and the ground. Some forest animals, such as elephants, ant-eaters and tapirs, spend their entire lives on the ground.

Most rainforest animals are small, so they can move freely through the trees. Large animals are very rare in the rainforest because food resources to sustain them are strictly limited. Dark conditions and few plants to feed on encourage few herbivores onto the forest floor, so carnivores face stiff competition for food supplies. Animals which eat plants derive their food supplies over a large area, because individual plants are widely scattered. As a result the forest usually appears to be relatively underpopulated with mammals, especially close to the ground.

Many forest animals of all sizes are well adapted to avoid being seen and eaten by predators, either through effective natural camouflaging of their appearance or by remaining stationary through the day and moving around in search of food at dusk or later.

The rainforest also provides a home for innumerable species of insects, which exploit whatever opportunities they can. For example, mosquitoes and other disease-carrying insects tend to feed on canopy-dwelling animals, only descending to the forest floor to disturb humans when the forest itself is disturbed or removed. Some animals and insects also play a vital role in pollinating rainforest plants. Ants play a critical role in destroying and recycling forest nutrients; some are vegetarian, but there are plenty of carnivorous ants which can strip a small animal within minutes.

1.5b Food webs

All forms of life within the rainforest are highly interdependent, so that even small changes in habitat or species can have serious knock-on effects throughout the ecosystem. If a food source (a species) is removed, the ecological chain reaction can be wide-ranging and long-lasting. Hence the serious concern in recent years over the stability and very survival of some rainforests which are threatened with irreversible change if not wholesale clearance.

Because of such variety of vegetation and bird, animal and insect life, food webs within the rainforest are diverse and complex, and therefore

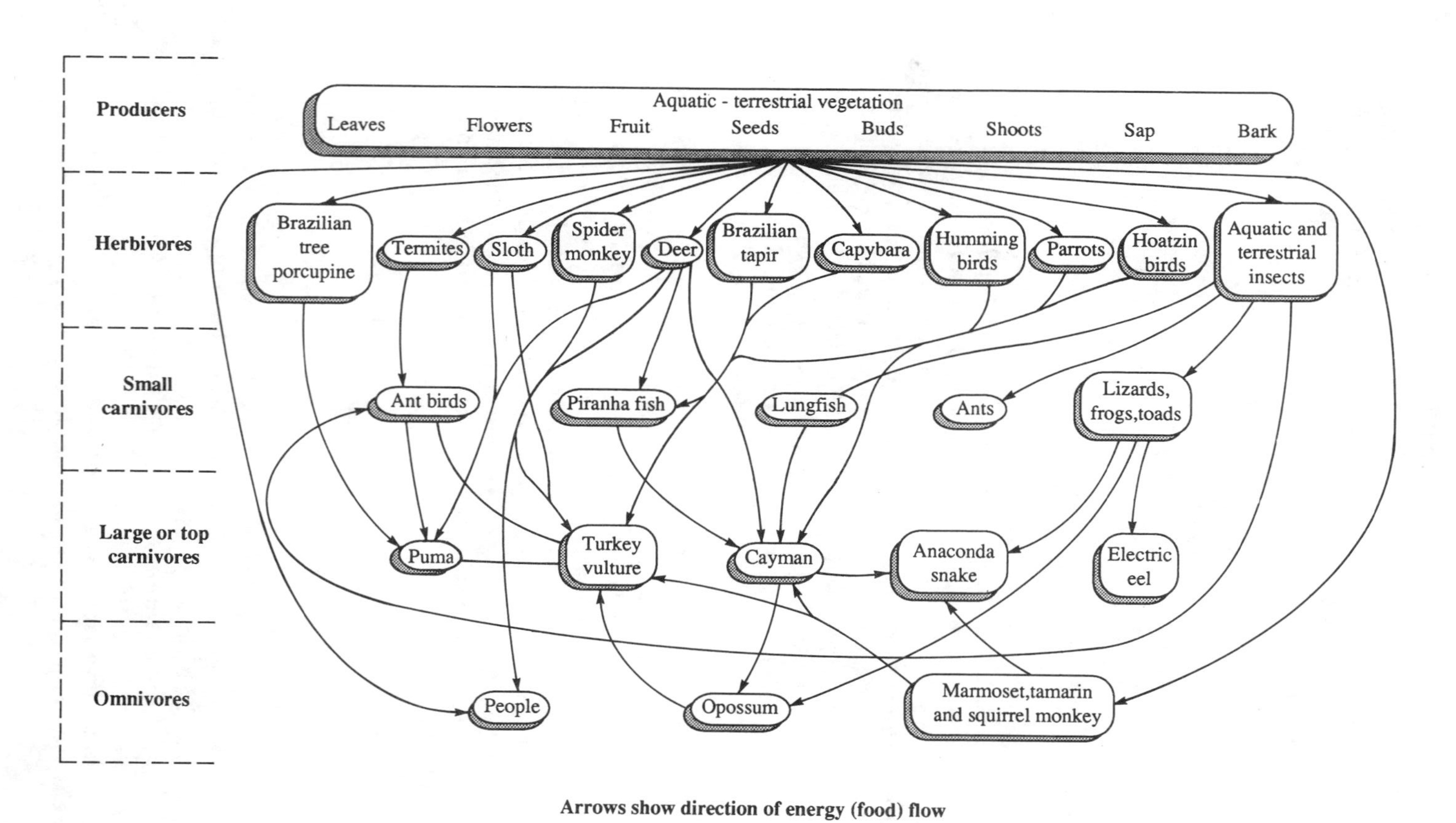

Figure 1.6 Brazilian tropical rainforest food web.

Source: after Law and Smith (1987)

difficult to investigate. An example is given in Figure 1.6, which illustrates the variety of primary food sources available to rainforest herbivores (grazers), which in turn are consumed by carnivores (meat-eating predators). The few rainforest omnivores (including people) have access to a range of food sources, and some may in turn fall prey to large carnivores.

The primary producers (plants) play a crucial role in rainforest dynamics. They absorb solar energy in the form of sunlight via their leaves and convert the energy into a usable form (sugars) by photosynthesis. Through transpiration they remove carbon dioxide from the atmosphere and replace it by oxygen, thus acting as an effective carbon sink in the global climatic system.

Consumption in the forest takes place most effectively and most efficiently in the canopy, where birds, bats and other animals feed on leaves and fruit and a few insectivores and mixed feeders are present too. Herbivores are the most common animals at all levels in the forest structure; carnivores are significant only in the upper canopy and on the ground.

When animals and plants die they fall to the forest floor. Decomposers quickly break down the remains and release the nutrients stored within them. Mineral nutrients are thus quickly recycled and returned to the rainforest soils, from where they can be taken in by plants and recirculated through the food chain remarkably quickly.

1.6 NUTRIENT CYCLING

Despite being generally underlain by old and inherently infertile soils, the tropical rainforest is a highly productive habitat. Scientists stress that 'the luxuriance of the vegetation disguises a basic infertility'.[30] The explanation of this curious paradox lies in the highly adapted forms of nutrient cycling which have evolved over long periods of stability in the rainforest. Unlike temperate forests, in which most of the nutrients released from decaying plant and animal remains are stored within soils and then made available to growing plants, most of the nutrients in the rainforests are locked up in the living tissues of the vegetation itself.[31]

1.6a Decay of organic remains

Some clues to this peculiar but highly efficient form of natural housekeeping are given in the general appearance of a rainforest. Rarely is it possible to find the decaying body of a dead animal; the remains are quickly broken down and no traces survive. The same is true of animal excrement and other forms of organic refuse. Leaves and flowers rain down from the tree canopy above in an almost unbroken stream in this world without seasons, yet they too quickly disappear. There is no thick mat of accumulated rotting vegetation on the forest floor, only a thin carpet of leaves. Trees and branches which fall onto the forest floor are quickly riddled by termites which digest them and break them down into dust with remarkable speed.

In the rainforest there are particularly rapid, effective and efficient ways in which the nutrients stored in living plants are recycled back to other plants, effectively short-circuiting the normal cycle in which storage within soils plays a vital role. Most involve organisms such as fungi, bacteria, worms, ants, moulds, termites and parasites which are encouraged to thrive within the rainforest by the prevailing environmental conditions (particularly humidity and temperature). These organisms quickly start to decay and break down dead organic matter (such as leaves, dead animals and fallen trees), so the stored nutrients are made readily available for re-use with minimum delay.

Much of the soluble mineral matter within the rainforest may never even reach the forest floor, because much of the falling leaf material and wood lands on the above-ground roots of trees and other plants. Many of these roots are covered with fungi (mycorrhizae) which rapidly break down the organic material and transfer the mineral nutrients directly back to the living plants via the finer roots themselves.

Insects also play important roles in the breakdown of organic material. Many of the rainforest insects, particularly the vast armies of leafcutter ants which maraud around the forest floor and over plants, deposit leaf material below ground and then cultivate mould on it to feed their young. This mould also serves to break down the leaves very quickly, releasing the nutrients within them for subsequent use by living plants.

1.6b Efficiency

Experimental studies demonstrate just how efficient the recycling of nutrients within the rainforest can be. In one such study 'tagged' calcium and phosphorus (labelled with radioactivity) were artificially added to the root mat of a patch of rainforest floor.[32] Rainwater was collected after it had drained through the plant roots, and it was found to contain less than 1 per cent of the labelled materials. The results suggest that nutrient recycling was over 99 per cent efficient, at least under the experimental conditions.

The highly efficient recycling of nutrients within the rainforest by organisms has two consequences – recycling is rapid, and only small quantities of mineral nutrients are available in the forest soils at any particular time (most are stored in the living vegetation). These are crucial aspects of ecosystem dynamics when considering management of the rainforests.

1.6c Management implications

Schemes to cut down rainforest and use the land for intensive farming are quite clearly misguided. They fail to recognise the natural integrity of rainforest nutrient cycles and the inherently unproductive and unsustainable character of most nutrient-poor rainforest soils. When forest is cleared, the natural community of soil fauna and flora is badly disrupted through exposure to sun and rain, lack of dead organic material to live off and the direct impacts of disturbance. This has critical effects on the recycling of nutrients.

The limited stock of nutrients is quickly leached out, the sun-baked soils develop hard impervious crusts, and valuable topsoil is eroded by raindrop impact and surface runoff. The net result is that the fertility of cleared areas is rarely sustained more than a few years, after which the soils quickly become barren and economically worthless.

Ironically, awareness that rainforest soils are not sustainable under agriculture is not new. The rapid depletion of soil nutrients after forest clearance has been known for at least four centuries, as the history of exploitation of the Brazilian rainforest reveals.[33] Portuguese colonists exhausted the supply of natural fibres and spices from the forests, and then tried to set up plantations. The plantations quickly failed because of soil depletion and exhaustion.

Yet the myth was created that rainforest soils are highly productive. It was spread by early explorers and naturalists like Alfred Wallace and Henry Bates in the mid-nineteenth century, who wrote of the rich potential of the Amazon rainforest to provide fertile pastures, gardens and orchards. Such fanciful speculations, and the survival of the myth, have done little to engender realistic attitudes towards the rainforests and their true potential.

1.7 RAINFOREST SOILS

Rainforest soils are very old compared to most other soils, having developed over long periods during which climate and environment have apparently been remarkably stable.

Variations in geology, topography, drainage and micro-climate can cause considerable spatial variability in tropical soils. Soil type and quality can change a great deal even over short distances, promoting a variable mosaic of natural vegetation and offering mixed prospects for productive farming after forest clearance.[34]

1.7a Soil character

Soil properties under the rainforest are quite distinct, and the soils differ from those in other environments in a number of ways.[35] They are usually of poor quality, with low mineral content. Ninety per cent of the soils in the Amazon Basin are deficient in nitrogen and phosphorus and 80 per cent are short on aluminium and potassium.[36]

Because up to 95 per cent of the nutrients required for plant growth are stored in the vegetation itself, rainforest soils are poor in macronutrients such as oxygen, carbon, hydrogen and nitrogen. They are generally acidic. On the positive side, these soils are generally well drained, with structures that offer good aeration. The upper horizon is often enriched with organic matter, making it a good environment for decomposition.

High temperatures and humidity promote deep weathering of bedrock, releasing clay-rich material. The rapid decomposition of organic matter means that bases and nutrients are returned to the soil quickly. Clay minerals break down and much of the silica and salts within the soils is washed away (leached). This leaves aluminium and iron oxides, which produce a red ferralitic soil (the iron may be hydrated in wet conditions to produce a yellow colour).

1.7b Laterite

Repeated upward and downward movement of the iron compounds under alternating wet and dry conditions can create a thick accumulation of iron compounds at the top of the water table. These will often build up into a hard impermeable layer called a laterite.

Laterites (oxisols and ultisols) are the most common rainforest soils. The upper soil horizon is quickly washed away, and the reddish-brown sheets of tough clay which remain are impermeable and create hard, pavement-like surfaces (especially when baked by the tropical sun). The name laterite is derived from the Latin word for brick precisely because these hard, dried and cracked surfaces resemble clay bricks.

1.7c Fertility

Although they support luxuriant vegetation, most tropical soils are not very fertile because long periods of weathering and leaching have washed the natural nutrients out of them. Up to 82 per cent of the tropical soils in the Americas, 56 per cent in Africa and 38 per cent in Asia are highly acidic and relatively infertile.[37] The most fertile tropical soils are found in areas where there is a regular supply of fresh mineral material, such as river floodplains and active volcanic areas.

These infertile rainforest soils have extremely limited potential for sustainable farming after forest clearance. Once the delicate natural balance between leaf fall, soil nutrients and plant growth is broken, the quick recycling of nutrients ceases and mineral deficiencies become acute.

1.7d Soil/vegetation interactions

Soils and vegetation in the rainforest are intimately linked. Inevitably soil type and character can exert strong control over the type and character of forest cover. One particularly significant set of interactions relates to the protective influence of forest cover on the underlying soil. The forest canopy protects soil from the impact of intense tropical rainfall, and the litter layer on the forest floor protects the soil from rain splash. Tree roots serve to bind the soil and thus reduce erosion potential.

The protective influence is quickly lost when forest cover is cleared. This makes the rainforest environment highly vulnerable, and many of the invaluable environmental services performed naturally by the forest are either lost or disturbed (see Chapter 4). Once forest cover has been removed, soils are very prone to leaching and erosion.[38]

Laterisation of soils – involving the washing away of the most fertile upper soil horizons and exposure of the hard laterite layer in which little if anything can grow – is a common consequence of rainforest clearance. This both limits the likelihood of subsequent sustained use of the soil, and inhibits any possible re-establishment of the forest cover.[39]

1.8 ECOLOGICAL DIVERSITY

Great species richness is probably the most distinctive characteristic of the tropical rainforests. Many early explorers who visited rainforests were struck by the large number of different species within them, and the unusual or exotic appearance of many of them. Alexander von Humbolt, a German scientist who explored Central and South America between 1799 and 1804, stressed the biological diversity of the region's rainforests. Darwin's description of the Brazilian rainforest around Rio de Janeiro (see page 1) spoke of 'the higher feelings of wonder, astonishment, and devotion, which fill and elevate the mind' when confronted with this wonderland of creation.

The rainforests are without doubt the most complex and energy-efficient ecosystems on earth, with a variety of species unrivalled anywhere else.[40] As such they are a great ecological asset, not only for the tropical countries they grow in. They are a natural resource of truly global value and significance.

The full extent of the rainforests' ecological diversity will never be known, and the race against time to complete the inventory before more species are lost through clearance adds urgency to the task.

The catalogue of known species is just the tip of a giant iceberg of unknown size because scientists have yet to discover, study, classify and name many rainforest species, especially of animals and insects. Birds

and mammals are the only two groups which have yet received adequate study.[41] Nature's vast storehouse of biological variety has untold potential for human use (such as in medicines, new crops and food sources), as well as incomparable ecological significance.

There is the added problem that many species found in the rainforest are not found elsewhere. These endemic species are particularly at risk from extinction caused by clear-felling or selective cutting of the forests.[42]

1.8a Global importance

The available statistics are impressive and there is no shortage of hyperboles to quote in support of the rainforest's claim to fame as the richest ecological zone on earth.[43] With more species of plant and animal than anywhere else, it is little wonder that the rainforests has been called 'the richest and most exuberant expression of life on land'.[44] They are also heralded as 'powerhouses of evolution'[45] because of their vast genetic diversity.

The global significance of the rainforests is striking. At least half of the known species of plants and animals in the world are found within the tropical rainforests, although they occupy only 8 per cent of the world's land area (see section 1.1). Over 60 per cent of all known species of plant (roughly 155,000 out of 250,000) are found in the tropical rainforests, along with 40 per cent of birds of prey and as many as 80 per cent of all known insects. About 90 per cent of the world's non-human primates (such as monkeys) are found only in tropical forests.[46] One in five of all the bird species on earth are believed to live in the Amazon rainforest.[47] An estimated 30 million species of insects live in the canopy above tropical forests.[48]

Naturally some rainforests have more species than others, because of environmental, evolutionary and genetic variability. The rainforests of South East Asia are believed to have the greatest density of species overall, and African rainforests are relatively species-poor in comparison.

Uncontested top of the list of rainforests in terms of species diversity is the Amazon in South America. Best estimates are all we have to judge by, but with possibly between 1.5 and 2 million species of plants and animals it has a greater variety of species than any other place on earth. The Amazon is home to an estimated 2,500 different tree species,[49] together with 2,000 described species of fish, 30,000 species of plant[50] and as many as 40,000 species of insects. It is estimated that the Amazon rainforest houses nearly half of the world total of 8,600 bird species (including 319 types of hummingbird alone, of which the whole of the USA has only 18).[51]

1.8b Species density

Perhaps the most impressive expression of species diversity is the density of species found in a given area of rainforest. Figures on the number of different species to be found within a given area of rainforest vary from study to study, but they are invariably high. Different studies use different unit areas and measurement scales; we can standardise a sample of findings using 1 hectare (10,000 m^2), roughly the size of one soccer pitch, as a basic unit of comparison.

Between 50 and 200 different tree species per hectare of rainforest seems typical,[52] and it is not uncommon to find well over 100 species belonging to more than 50 genera within each 1 ha of rainforest.[53] Diversity can be even higher in unusually rich pockets of rainforest; a 100 m^2 (1 per cent of a hectare) patch of tropical rainforest may contain up to 230 different tree species.

A similar area (1 ha) in a typical species-poor temperate forest might contain between 7 and 10 tree species.[54] One hectare of North American deciduous forest might have between 10 and 30 tree species, and coniferous forests in the far north of Canada might have between 1 and 5 species.[55]

This huge species diversity is not confined to trees. In fact larger trees make up only a small percentage of the total diversity of rainforests. There are many more species of herbaceous plants including ferns and epiphytes, as well as many less complex plants such as lichens and fungi. It is estimated that each species of rainforest tree supports over 400 unique insect species. A recent study of one hectare of Peruvian rainforest found 41,000 different species of insects living in the tree canopy (including 12,000 species of beetle).[56]

The great diversity of species of all types within a given rainforest means considerable competition between species and thus a lack of dominant species. This is true at all levels in the rainforest food web, from trees through grazing species to carnivores. It gives variety to the appearance of the forest, and contributes to the impression of luxuriance and abundance noted by many observers. The lack of dominant species also means that individual examples of any particular species are often hard to find, because they are widely scattered through the forest in small numbers. Even tree species in the rainforest are widely dispersed, not closely grouped in stands as in natural temperate forests. It is estimated, for example, that natural primary Malaysian rainforest may contain up to 250,000 individual plants per hectare.

1.8c International comparisons

Some interesting international comparisons are often drawn to illustrate just how high the species diversity in rainforests really is. For example, the 41,000 species of insects in 1 ha of Peruvian rainforest compares with a total of 1,430 insect species in the whole of Great Britain.[57] The United Kingdom has 1,443 named species of plants, but because of their species-rich rainforests Costa Rica (a fifth of the size) has over 8,000 and New Caledonia in the Pacific (a twentieth of the size of Britain) has 3,000 species.[58]

Ecuador, which is slightly bigger than Britain, has an estimated 20–25,000 plant species (up to 5,000 of which are endemic), including up to 3,000 species of orchid alone.[59] Britain has around 50 species and there are 153 in the whole of North America. Ecuador also has over 1,550 species of bird (17 per cent of the world total), 280 species of mammal, 345 species of reptiles and 358 of amphibians. And these are just the known and named species.

Tropical rainforests often have between 5 and 20 times more tree species than temperate forests. Canada and the USA together have a total of around 700 different tree species, but this number can be found within ten 1 hectare plots in the rainforests of Borneo. The tropical island of Madagascar alone has 2,000 tree species.[60]

1.8d Causes of diversity

There is little general agreement over why species diversity in the rainforest is so high, and many different reasons are suggested.[61] Diversity has clearly been created by a number of factors peculiar to the tropical rainforest environment, which include the following.

The great age of the rainforest system is doubtless very important. But perhaps even more important is the long period of climatic stability (estimated to be at least 75 million years without major disturbance), especially when compared with the northern temperate zone. Longevity and stability mean that rainforest plants and animals have had time to evolve more or less without interruptions, during which they have been able to adapt to produce optimum variety.

Geographical factors might also have conspired to evolve diversity, because many rainforest areas may have enjoyed enough geographical isolation over the recent geological past to allow the evolution of completely new species. Precisely how this local speciation arises is open to debate. It might be a product of natural selection (Darwin's 'survival of the fittest'), or it might arise through random genetic drift. Rainforests doubtless offered invaluable ecological refuges to some species displaced from higher latitudes by long-term environmental (particularly climatic) changes, which might have promoted evolutionary divergence.

Certainly the rainforest environment has some valuable evolutionary advantages over others. Optimum growth conditions are provided for many species by the rainforest climate and soils. Growth throughout the year, encouraged by the lack of seasonality, allows constant reproductive ability. One result of this is that processes of natural selection can act at all times, and thus evolutionary change can be continuous and cumulative. Thus the pace of evolution can be unusually fast in the rainforest.

Advantages also accrue from the great variety of ecological niches available within the rainforest, which allows greater speciation and less competition between species. Niches in the rainforest – determined largely by habitat and food sources – tend to be ecologically narrow (because there are so many of them). The organisms which occupy them are mostly specialists with precise adaptations to their environment, which perform specialised if not unique functions. Interdependence between species in the rainforest is common (for example, particular insects assist in the pollination of particular plants). Indeed such mutually beneficial co-operation between species is essential to the stability of the complex rainforest system.

Plant–animal interactions in the rainforest also encourage high species diversity. The many different species of trees and other plants provide a varied range of habitats and food sources, which encourages many species of birds, insects and animals. But the relationships are often symbiotic (two-way benefits) because the birds, insects and animals aid seed distribution and promote dispersion of plant species. Some biologists account for the lack of dominance by any single species of plant or tree as the result of herbivores eating seedlings and saplings, thus preventing large populations of particular species.

A large species diversity in the rainforest might also reflect the varying mosaic of habitats and ecological niches caused by environmental disruption. This 'intermediate disturbance hypothesis' argues that natural disruptions caused by floods, droughts, natural forest fires and even tectonic activity have played a part in encouraging the evolution of a large number of different species.[62] Such diversity of habitat can also be caused by traditional patterns of land use by native forest peoples, involving shifting cultivation (see page 46). The net result is a patchwork mosaic of vegetation at different stages in this natural self-repair cycle, which creates a variety of habitats and encourages more different species into the area of rainforest.

1.9 VALUES OF THE TROPICAL RAINFOREST

Whatever the reason or reasons for their striking species diversity, there is no doubting the fact that tropical rainforests are special. But why do we care about them and their survival? There are, basically, two sets of reasons – intrinsic and extrinsic.

The intrinsic value of the rainforests means that we care, quite simply, because they are there. The argument is basically a moral one, a point of principle which transcends any self-interest we might personally have in the forests. Aesthetic interest is deep-rooted. We appreciate the beauty, abundance and variety of the natural rainforest environment which is more or less untouched by human activities.

The very fact that it is there matters; we may never actually visit the rainforest to experience this aesthetic benefit in person. Ethical interest also figures large. The ethics of conservation argue that rainforests must be protected because they are home to wildlife which has a right to be left alone. The ethics of humanitarianism argue that rainforests are also home to an estimated 50 million tribal people who also have a right to be left alone (see Chapter 5).

The extrinsic values of the rainforests mean that we care because they meet some of our material needs. This utilitarian argument (see Chapter 4) has both economic and functional dimensions. Nature's warehouse of raw materials provides rich yields of forest products which are used in medicine, agriculture and industry, as well as tropical timber. The forests also provide some critical environmental services, such as limiting soil erosion and downstream silting, controlling drainage basin hydrological processes and thus downstream flooding, and influencing climate on the local, regional and even global scales.

At least one argument combines the intrinsic and extrinsic dimensions. We care because we appear to be throwing away the unknown by clearing rainforests and promoting wholesale extinction of species. Inevitably the extinctions include many undiscovered species; we are destroying them before we even know they exist.

The rainforest clearance debate raises very important questions about how we value natural resources, and what form of stewardship is most appropriate to ensure that we use them sustainably.[63]

So, the costs of losing the remaining rainforests are wide-ranging and the stakes may be very high indeed. In Chapter 2 we review the evidence about rates and patterns of forest clearance, and in Chapter 3 we examine the causes of clearance.

2

DESTRUCTION OF THE RAINFOREST: RATES OF LOSS

Cut is the branch that might have grown full straight.
Christopher Marlowe (1564–93), *Doctor Faustus*

2.1 INTRODUCTION

The rainforests are under attack. These rich and complex ecosystems, which have survived millions of years of natural environmental change (indeed they have flourished through it), are now facing a fight for survival. The hands of people are inflicting more damage on the rainforests in a matter of years than the entire forces of nature have done over geological time-scales.

Norman Myers, an international expert on rainforests, pointed out early in 1990 that 'at issue is the most exuberant expressions of nature that has ever graced the face of the planet during four billion years of evolution. Within just another 40 years at most, we may see the last remnants fall to the chainsaw and the matchbox.'[1] The timetable is open to debate; that the fight for survival is on is not.

2.1a Shrinking forests

We have already seen some evidence in Chapter 1 that today's rainforests are shrunken remnants of much larger forests from the ancient past. These survivors represent the outcome of long periods of climatic change; they are natural distributions, in equilibrium with today's climatic constraints in the tropics.

But even that picture reflects a theoretical distribution rather than an actual pattern of vegetation on the ground. The maps of world vegetation distribution (Figures 1.1 and 1.2), for example, show climatic climax vegetation – what *should* exist under prevailing climate, in the absence of damaging human activities, rather than what *does* exist. There is little doubt that many areas shown on the maps as rainforest no longer have natural forest cover, having been cleared for one reason or another.

Plate 1 Rainforest clearance in Amazonia, Brazil. After commercially valuable trees have been removed through selective logging, the area has been abandoned and is now waste land.

Source: H. Girardet/The Environmental Picture Library

Disparities between theoretical and actual distributions of rainforest reflect human disturbance of the forest habitat. This comes in two forms. Degradation involves complete loss of the forest, which might be cut down and replaced by open woodland or agriculture. The loss is permanent. Depletion involves some change to the forest ecosystem, but not complete removal. Some plant and animal species are lost, but forest remains (albeit a much-modified forest). Natural regeneration can re-establish the forest ecosystem, given a long enough period without further depletion. Both forms of disturbance of rainforests are widespread, but degradation poses the greatest threat.

2.2 PAST RATES OF CLEARANCE

Precise figures of the amount of rainforest which has been lost as a result of human activities are elusive because we can only make informed guesses about what the original natural distribution of rainforest might have been (based largely on reconstructions of past climates). Estimates vary. Some studies suggest that tropical rainforest might once have covered an area of up to 15 million km^2 (about a tenth of the earth's land surface). Estimates of the amount of destruction caused by human activities vary between a third and half of the original.[2]

2.2a Early clearance

What is more certain is that clearance of the rainforest has been going on for a long time.[3] There is evidence of clearance for agriculture at least 3,000 years ago in Africa, 7,000 years ago in South and Central America and possibly 9,000 years ago in India and New Guinea.[4] Traditional forms of forest clearance by burning were small-scale and localised and they had relatively little impact on the overall extent, distribution and character of the rainforests. Indeed they may even have contributed to development of the diversity of species (see section 1.8).

More recent exploration of the rainforests, prompted by the search for commercially useful resources as well as by land hunger, started the irreversible tide of forest destruction and clearance. Early episodes were small-scale and isolated. During the fifteenth century, for example, groups of English and Dutch migrants lured by a gold rush looked to the Brazilian Amazon to meet their need for food and charcoal. Forest species were exploited for food; trees were felled and burned for charcoal. In the eighteenth century parcels of rainforest were cleared from the hills of central Minas, in eastern Brazil, to create land for cattle ranching. Soil depletion and erosion quickly followed.

More widespread exploitation of the rainforests began during the eighteenth and nineteenth centuries as demand started to grow in the western world for tropical plantation crops.

2.2b Historic rates

Estimates of long-term rates of clearance vary considerably, and it is difficult to arrive at a universally agreed figure. Table 2.1 shows a recent estimate of the original and present areas of tropical forest compared with other major ecosystems. The data suggest that clearance of the tropical forest (including but not confined to rainforest) has been much more limited than changes in other types of forest and woodland, and changes in grassland. But these data appear seriously to under-represent rainforest clearance; recent data suggest that, overall, the tropical rainforests have declined by perhaps half.

Table 2.1 Estimate of pre-agricultural and present area of major biomes

	(Areas in million km²)		
Biome	*Pre-agricultural*	*Present*	*% change*
Tropical rainforest	12.77	12.29	–3.8
Other forest	33.51	26.98	–19.5
Woodland	15.23	13.10	–14.0
Shrubland	12.99	12.12	–6.7
Grassland	33.90	27.43	–19.1
Tundra	7.34	7.34	0
Desert	15.82	15.57	–1.6
Cultivation	0.93	17.56	+1,788.2

Source: Matthews (1983).

Deforestation of rainforests (involving both degradation and depletion) has accelerated significantly since the turn of the century, and particularly since 1945. What is new is the speed and distribution of clearance. In the past it was relatively slow and limited in spatial extent; now it is often fast and certainly widespread.

UN Food and Agriculture Organisation (FAO) figures suggest that half the world's forests have disappeared since 1950.[5] Losses have been variable but high within the main forested areas. The biggest relative losses have been in Central America (66 per cent) and Central Africa (52 per cent), although Latin America and South-East Asia have both lost over a third (37 per cent and 38 per cent respectively).[6] Some of the specific casualties of rainforest clearance are listed in Table 2.2.

Table 2.2 Extent of tropical rainforest loss, to 1988

Losses include:	
Bangladesh	all primary rainforest lost by 1988
China	50% forest loss in Xishuangbana Province
Haiti	all primary rainforest lost by 1988
India	all primary rainforest lost by 1988
Philippines	55% forest loss 1960–85
Sri Lanka	almost all primary rainforest lost by 1988
Thailand	45% forest loss 1961–85
WORLD TOTAL	over 40% tropical forest lost by 1988

Source: summarised from World Wide Fund for Nature (1988).

2.3 PRESENT RATES OF CLEARANCE

2.3a Reliability of evidence

Reliable estimates of rates and patterns of clearance have until recently been difficult to secure, and there has been considerable disagreement over the methodologies used and estimates produced.[7]

Trying to compare results from different studies is fraught with difficulties. Some observations are clearly more reliable than others, there is no universally agreed method of measuring and calculating deforestation rates, and different time scales are used. Some countries have no baseline surveys against which to compare recent changes.

Definitions vary, too. Some figures include degradation and depletion, while others only cover the former; some studies include all tropical forests, whereas others only relate to rainforests. Up-to-date survey data are simply not available for many regions. It is not inconceivable that some countries do not make available all of their deforestation records, in an attempt to hide the true extent of their problems.

2.3b Recent estimates

Estimates of annual rates of rainforest clearance vary considerably. Figures from FAO studies in the early 1980s (Table 2.3) suggest total annual clearance of about 113,000 km^2 a year, roughly three-quarters of which was closed rainforest. Rates of rainforest clearance were higher in absolute terms in the Americas, but broadly comparable for the three areas in terms of percentage of remaining forest cover.

Where detailed evidence is available of recent changes, it suggests that deforestation rates have continued to rise. Over half of the total forest clearance in Brazil up to the year 1978 is believed to have occurred in the three years 1975 to 1978,[8] for example.

The most recent measurements, from the late 1980s, show that

Table 2.3 Estimated rates of tropical forest clearance, to early 1980s

	(Area in thousands of km²)					
	Closed forests		*Open forests*		*Total*	
Region	*Area*	*% clearance rate*	*Area*	*% clearance rate*	*Area*	*% clearance rate*
Africa	13.3	0.61	23.5	0.48	36.8	0.52
Americas	43.4	0.64	12.7	0.59	56.1	0.63
Asia	18.2	0.60	1.9	0.61	20.1	0.60
TOTAL	74.9	0.62	38.1	0.52	113.0	0.58

Source: Food and Agriculture Organisation (1982).

tropical forests world-wide are being cleared at a rate of around 100,000 km² a year.[9] Roughly 60 per cent of the clearance is believed to be rainforest; the rest is seasonally wet and dry forms of tropical forest.[10]

A further 100,000 km² of tropical forest is seriously damaged or partially cleared each year.[11] Partial clearance is often associated with logging (see section 3.6) and extensive slash-and-burn farming (see section 3.3); its ecological and environmental impacts are often as serious as those arising from total clearance.

An area of tropical forest roughly the size of England, Scotland and Wales (229,878 km²) is cleared – totally or partially – every year[12] – nearly 4,000 km² a week or 38 ha every minute.

Estimates of the rates of rainforest clearance vary a great deal. Recently published figures range from 75,000 km² a year estimated by the UN Food and Agriculture Organisation,[13] through the 130,000 km² per year estimated by the Rainforest Foundation[14] to the 200,000 km² a year maximum estimate suggested by the US National Academy of Sciences.[15]

The 61,000 km² annual clearance of rainforest in the late 1980s, proposed by the World Resources Institute[16] and others, seems a reasonable figure to accept as a benchmark.

2.3c Remote sensing and data collection

Recent studies show that tropical deforestation may be much more serious than previously thought,[17] but the recent estimates may include deforestation previously not measured or calculated, as well as recent clearance. It is difficult to establish how much of the reported increases in deforestation rates might reflect improved methods of measurement and how much reflects a real increase in the area being cleared. Doubtless both are occurring.

Remote sensing, particularly using satellites, holds much promise for revealing exactly where and when deforestation is taking place.[18] Yet the promise is being realised only slowly and the available technology has not been universally applied mainly because of high costs.

Where satellite surveillance and improved photographic interpretation have been used, the results are startling. They indicate that earlier statistics have been open to many errors and distortions. Satellite imagery has often established that government figures are unreliable. For example, in the early 1980s the Philippines government claimed that 58 per cent of its land was covered by tropical forest, but satellite imagery showed the true figure to be nearer 38 per cent.[19] Satellite imagery was used in a 1987 study for Brazil's National Space Research Institute (INPE) to determine that 80,000 km^2 of virgin forest in the Amazon Basin had been cleared that year (this figure was later reduced because it included second growth clearing). Other INPE remote sensing surveys showed total deforestation in the Amazon Basin of 170,000 km^2 over the period 1978–88.

Where remote sensing information is available for the same area at different points in time, it is possible to see clearly where and how fast forest clearance is progressing. Two such sets of time-series, from Rondonia in Brazil (Figure 2.1), show rapid clearance of patches of forest by settlers in the late 1970s, with deforestation concentrated along the main highways and feeder roads.

Better quality information derived from remote sensing and satellite imagery suggests the overall rate of loss of rainforests (204,000 km^2 a year) is over 80 per cent higher than the estimates by the UN Food and Agriculture Organisation in 1980 (114,000 km^2 a year).[20] Yet the FAO estimates have been the basis of much political and conservation decision-making. Some particular estimates are proving to have been wrong by orders of magnitude. For example, the rate of clearance in India is now estimated at 150,000 km^2 a year, nearly ten times the 1980 FAO estimate of 1,470 km^2 a year.[21]

Whilst sophisticated remote sensing technologies are a considerable improvement over more traditional approaches, based mainly on ground surveys, they are not without problems and limitations. Satellite imagery can be inaccurate because some types of imagery (such as VHRR) only detect the amount of smoke over an area. Deforestation without fire can remain undetected (giving an underestimate), while smoke from forest fires can drift a long way downwind from the actual fire (giving an overestimate). Not all satellite imagery suffers from these drawbacks, however. Landsat yields good estimates of deforestation. Remote sensing using a mixture of ground surveys and airborne sampling methods also has its problems. For example, only broad classes of vegetation can be distinguished, airborne surveys are liable to errors caused by cloudiness and smoke from fires, and fine resolution requires detailed surveying which can be prohibitively expensive.

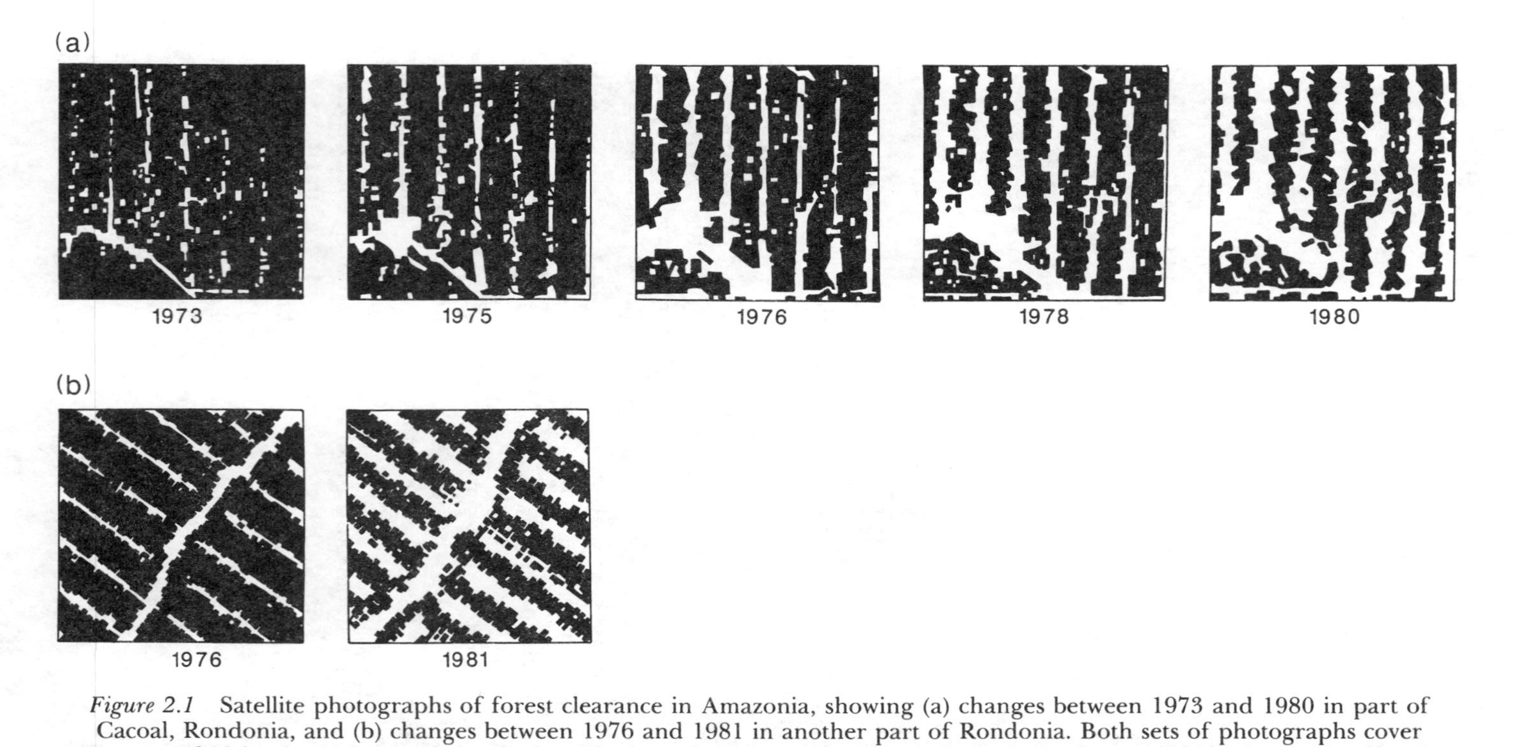

Figure 2.1 Satellite photographs of forest clearance in Amazonia, showing (a) changes between 1973 and 1980 in part of Cacoal, Rondonia, and (b) changes between 1976 and 1981 in another part of Rondonia. Both sets of photographs cover an area of 30 km by 30 km, and clearly show the impact of the road network through the forest, with feeder roads off at 5 km intervals. Colonists were provided by the government with plots 2,000 metres long by 500 metres wide, which are the basic units of clearance.

Sources: (a) after Johnson *et al.* (1989b), (b) after Webster and Williams (1988)

2.4 PRESENT PATTERNS OF CLEARANCE

Recent surveys show that clearance is occurring in all areas where significant remnants of rainforest remain, but that rates, causes and consequences of deforestation are not uniform. Some countries face relatively few problems. Present rates of clearance in Venezuela, for example, are negligible. But other countries face massive losses. By the end of the 1980s, for example, over 80 per cent of the rainforest had been cleared in Nigeria, and nearly 90 per cent had been cleared in Bolivia.

2.4a Hot spots

The main 'hot spot' areas, in which rainforests are critically threatened, are distributed unevenly between the continents (Figure 2.2). Three large areas in South America are on the critical list – the uplands of western Amazonia, the Atlantic coast of Brazil and in western Ecuador. Amazonia, in particular, faces serious threats.[22] Rainforests in Madagascar, off the east coast of Africa, are also critically threatened. In the Far East[23] the main threats are in peninsular Malaysia, Indonesia, the Philippines, parts of the eastern Himalayas, Queensland in north-east Australia, and New Caledonia.

The ten leading rainforest hot spots account for about 3.5 per cent of the world's remaining tropical forests (292,000 km^2).[24] Many of them have unusually high species diversity and are under imminent threat of complete destruction. Estimates of the number of species at risk in them vary; some studies forecast that up to 17,000 plant species and at least 350,000 animal species may become extinct during the 1990s.[25]

2.4b Amazonia

Brazil is the biggest deforester (in terms of area and speed), accounting for about three-quarters of total world rainforest clearance. The Amazon rainforest in Brazil covers an area of 3.37 million km^2, and an estimated 148,000 km^2 had been cleared by 1983. Roughly 10,000 km^2 more is cleared each year.

The Amazon figures are having to be revised as better information comes to light which shows the situation to be much worse than previously imagined. Satellite data from Brazil's National Space Research Institute show that during the 1987 burning season (August to October) nearly 80,000 km^2 of forest was devastated by fire. Almost all the fires were illegal and by sides of roads.[26] The last year that land credits were available, 1987, seems to be the historic peak. Tax credits were subsequently suspended and eventually cancelled. Since then there has

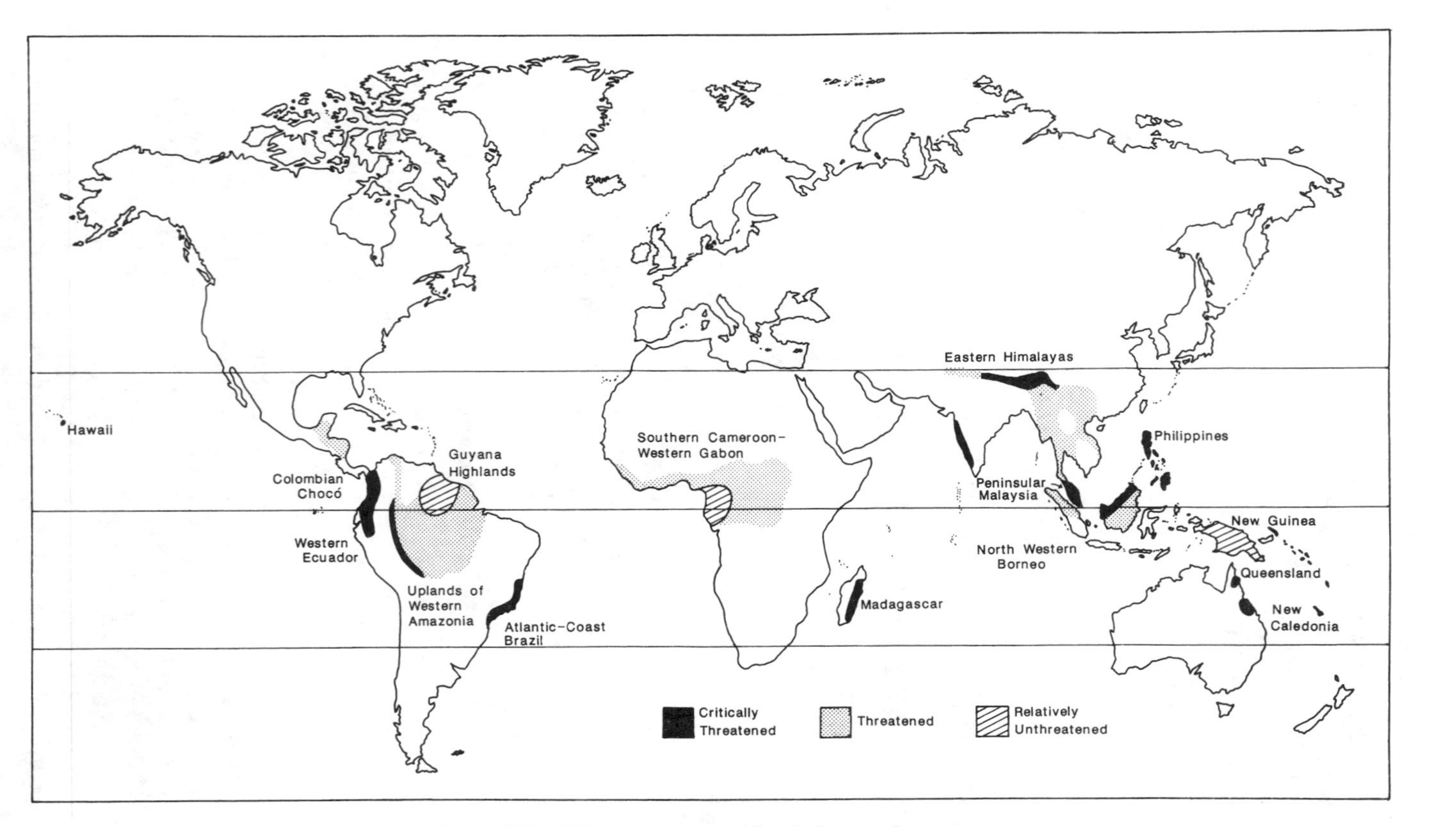

Figure 2.2 'Hot spot' areas of rainforest clearance.

Source: after Myers (1988a)

Table 2.4 Loss of tropical rainforests by country or region

Country or region	*Estimated date of total clearance*
(1) Rapid removal	
Australia, lowland Indonesia and Malaysia, Philippines, Thailand, Brazil, Central America, Madagascar, East Africa, West Africa	pre-1990
Bangladesh, India, Melanesia, Sri Lanka, Vietnam, Ecuador	by 1990
Remote Indonesia and Malaysia, Colombia	by 2000
(2) Remote removal	
Burma, Papua New Guinea, Peru, Cameroon	by 2000
(3) Currently showing slow change	
French Guiana, Guyana and Surinam, Zaïre Basin	post-2000

Source: Myers (1984).

been a slow-down in the rate of felling. A 40 per cent drop was recorded in 1988, and the rate fell by about half again in 1989. This trend has not been mirrored in other countries, however.

Some nations face the real possibility of losing all their tropical rainforests within the foreseeable future, particularly where rates of clearance are high. Clearance is proceeding at about 5.2 per cent a year in the Ivory Coast, for example. Costa Rica is losing its rainforests at a rate of about 3.5 per cent per year, Sri Lanka at 3.5 per cent per year and El Salvador at an estimated 3.2 per cent per year.

Rates of clearance are likely to accelerate in the future, if anything, in most areas.

2.5 FUTURE PROSPECTS

Such stark statistics on rates of rainforest clearance have given rise to some bleak forecasts about what the future holds in store. Reliable forecasts from the mid-1980s, using the best information then available, were suggesting that about 40 per cent of the remaining forest cover would be lost by the year 2000.[27] Well-intended forecasts do sometimes turn out to be wrong. For example, many of the rainforests predicted to have disappeared by 1990 (Table 2.4) were still there, but smaller.

But better monitoring since the mid-1980s has indicated much faster clearance and shorter life expectancy. Recent estimates of the length of time remaining for the world's surviving rainforests range from 20 to 40 years depending on the source.[28]

Predictions by the World Wide Fund for Nature[29] suggest that, at today's rates of clearance, most existing rainforests will be badly

Table 2.5 Forecasts of likely future extent of tropical rainforest loss, to 2000

Likely future losses predicted by the year 2000, based on proportions of the remaining forests:

Brazil	one-third lost
Colombia	one-third lost
Congo	two-thirds scheduled for logging
Costa Rica	four-fifths lost
Ecuador	over half lost
Ghana	over a quarter lost
Guatemala	one-third lost
Guinea	one-third lost
Honduras	over half lost
Indonesia	a tenth lost
Ivory Coast	almost all logged out
Madagascar	nearly a third lost
Malaysia	a quarter lost
Mexico	one-third lost
Nicaragua	over half lost
Nigeria	all lost
Peninsular Malaysia	all lost
Philippines	a fifth lost
Thailand	nearly two-thirds lost

Source: summarised from World Wide Fund for Nature (1988).

depleted by the year 2000 (Table 2.5) and all tropical forests outside protected areas will be seriously damaged by 2020. They also forecast that there will be no undamaged tropical forests of any sort by the year 2070, and conclude that 'what took millions of years to evolve will be lost during the lifetimes of children born today'.[30]

Not all studies are so *optimistic*! Many ecologists believe that, if clearance continues at present rates, only two large remnants of tropical rainforest are likely to remain by the year 2010 – western Amazonia and an area centred around central Zaïre.[31] Even these, it is argued, will not survive much past the turn of the century.

In Chapter 3 we examine the causes and processes of clearance.

3

CAUSES AND PROCESSES OF CLEARANCE

A fool sees not the same tree that a wise man sees.
William Blake (1757–1827), *Proverbs of Hell*

3.1 INTRODUCTION

In this chapter we examine what is causing the clearance of the rainforests. The glib answer sometimes offered to the question 'Why are rainforests under attack?' is simply that there are too many people expecting too many things from them. To a large degree this is true, but the reality is neither as simple nor as direct as the answer suggests. To isolate population as *the* root cause of forest clearance doesn't throw much light on what the main sources of pressure are, where the pressure is being applied most forcefully and where the pressures originate. Rainforest are being cleared for many different reasons.[1]

3.1a Pressure and people

To say that clearance is caused by over-population, high population pressure and rapid population growth in the developing countries of the tropics is not particularly helpful either. The available statistics *do* indicate high population growth rates there (Table 3.1).

But the demographic argument is too simplistic to have real merit, for at least two reasons. First, it fails to recognise that different forces are at work in different rainforests. To speak of deforestation as if it were a single process is misleading without examining who is clearing the trees and why. Second, it fails to recognise that much of the pressure on the rainforests originates from beyond the tropics, often in the developed nations of the north. Examples include the international demand for tropical hardwood and cheap beef, and the readiness in the past of governments and agencies (such as the World Bank) in the developed countries to grant financial aid for environmentally damaging development projects.

Table 3.1 Population projections for major regions

Region or country	*1975 (millions)*	*2000*	*% increase by 2000*	*average annual % increase*
Africa	399	814	104	2.9
Asia/Oceania	2,274	3,630	60	1.9
Latin America	325	637	96	2.7
Brazil	109	226	108	2.9
Indonesia	135	226	68	2.1

Source: Barney (1980).

There are three main links between population pressure and rainforest clearance in developing tropical countries – through the continued collection and use of fuelwood (see section 3.2), the progressive intensification of traditional (shifting) farming techniques (see section 3.3) and government-led resettlement schemes (see section 3.5).

But population pressure is only one of the three main forces of rainforest clearance. Rainforest trees are a valuable hardwood timber resource which has great economic value, particularly in developed countries. As a consequence commercial logging is directly responsible for much forest clearance (see section 3.6). The third purpose is to create grazing land for extensive ranching, particularly of beef cattle for the lucrative export market (see section 3.8).

There is an additional set of pressures which arise from the perceived need for economic development in many developing tropical countries and are manifest through major development projects (such as the construction of hydroelectric schemes and major highways) (see section 3.9).

3.1b Problems and patterns

It is difficult to judge how important one pressure is relative to another, particularly on a global scale. The figures in Table 3.2 offer one view, which suggests that by far the major culprit is farming activities (intensification of traditional practices). But the data should be treated with caution because they are based on very generalised estimates, they relate to the early 1980s (since when rates of clearance have increased greatly, and the balance of causes may well have altered too), and they make no mention of development projects.

The main source and the scale of pressure to clear rainforests vary a great deal from one country to another, and to a lesser extent between forests within a given country. By the late 1980s the main pressures in Asia were from commercial logging and agricultural expansion. In Africa the pressures have shifted away from logging, partly because

Table 3.2 Estimated rates of clearance of tropical rainforests for different purposes

Purpose	*Annual loss (km²)*	%
Commercial timber removal	45,000	18
Fuelwood gathering	25,000	10
Cattle grazing (South America)	20,000	8
Farming operations (minimal estate)	160,000	64
TOTAL	250,000	100

Source: Myers (1984).

many countries (like the Ivory Coast, Nigeria and large parts of the Congo) are logged out or nearly so. The main threats in Africa are now from cutting for fuelwood and overgrazing of cattle. Latin America has the largest remaining area of uncut forest and faces the widest range of threats, although commercial logging is limited. Most rainforest clearance there is associated with cattle ranching, population resettlement schemes and major development projects.

3.2 FUELWOOD GATHERING

The cutting and collection of fuelwood (firewood) by people living near the rainforest are a major cause of forest clearance in some areas, particularly in drier tropical forests.[2] It is estimated that in developing countries about 80 per cent of the wood used is for fuel; the rest is used by industry (the ratio is roughly reversed in developed countries).[3]

This local, sustainable energy supply is traditional and has obvious advantages – it is free, easy to collect and light to carry. Often it is the only source of fuel people have access to. Fuelwood shortages bring serious problems for those who rely on it.[4] Animal dung, which is really needed to fertilise crops, must be burned instead. Sparing use of available wood means that often a meal can only be cooked every few days. Malnutrition can follow, and disease spreads faster when food and water are not heated.

3.2a Scale and problems

The amount of wood collected is significant, and it has risen sharply in recent decades. Estimates are inevitably open to considerable margins of error, but the World Resources Institute[5] believes that the total annual production of fuelwood and charcoal in developing countries has doubled from less than 600 million m^3 in 1963 to more than 1,300 million m^3 in 1983. Fuelwood accounts for much of the increase.

The consequences of fuelwood gathering are directly related to population pressure. Under low population pressure gathering can be sustainable (rates of regrowth are higher than rates of clearance). Indeed, traditionally it has been. But increasing population pressure brings greater demand for wood, so more is gathered and it is required more regularly. Wood is harvested faster than it can naturally grow back, resulting in net clearance of rainforest.

The collection of fuelwood normally threatens the forest margins first. Once these accessible areas have been cleared or over-collected, gatherers must travel further into the forest interior in search of adequate wood supplies. Intensive fuelwood gathering in rural areas has been likened to 'mining of forest for fuel'.[6]

The United Nations estimates that 1.5 billion of the 2 billion people around the world who rely on fuelwood for cooking and heating are cutting wood faster than it is able to grow back naturally. An estimated 125 million people in 23 countries – 60 per cent of the population that depends on fuelwood for cooking and heating – cannot get enough fuelwood, even by over-cutting forests.[7] The problem is particularly serious in Africa, where up to 180 million people faced acute shortages of firewood in 1980.

3.2b Alternatives

Logically these fuelwood users ought to switch to other energy sources to alleviate the increasing journeys involved in finding usable supplies, reduce their dependence on natural resources which are proving to be unsustainable at present rates of use, and minimise their damaging impacts on the rainforests. But they face severe if not insurmountable constraints – of tradition, poverty, lack of available technology and lack of information about viable alternatives.

There are ways of relieving some of the pressures on forests from fuelwood gathering, but they need to be much more widely adopted. Village woodlots could be planted to provide a convenient, accessible and sustainable yield of wood for all purposes. Greater use could be made of wood from tree species which burn best, rather than indiscriminately taking all the wood from within easy walking distance. Metal cookers could be replaced by clay ones which need half as much wood to heat (they are also cheap and time-saving to use).

Problems of fuelwood shortages and related increases in forest clearing are likely to get worse rather than better in the foreseeable future, because replanting is woefully inadequate. It is estimated that less than a tenth of the area cleared for fuelwood is replanted.

3.3 SHIFTING CULTIVATION

Although the natural rainforest is a luxurious, highly varied and highly productive type of vegetation, the underlying soils are inherently infertile (see section 1.7). Yet for generations forest peoples have been able to exploit their environment in ways which are sustainable, bring no long-term damage to the forest and protect soils from erosion and declining fertility.[8]

Their practices are generally referred to as shifting cultivation, because they are based on periodic abandonment of worked areas and clearance of new patches of forest. The more traditional name, 'slash and burn', describes how the patches in the forest are cleared to create land to grow crops on.

3.3a Rates and patterns

Peasant farmers, with their shifting cultivation, are often blamed as the major cause of tropical deforestation. The evidence, patchy as it is, does suggest that forest clearance for farming far outstrips other causes (Table 3.2). More recent estimates suggest that shifting cultivation activities destroy 50,000 km^2 and degrade a further 10 million km^2 of tropical rainforest a year.[9]

Whilst each farmer works a small patch and usually moves on every 4 to 8 years, there are many of them and the net effect is cumulative. The UN Food and Agriculture Organisation estimates that something like 250 million shifting cultivators are using slash-and-burn techniques to colonise rainforests.[10]

Forest clearance by shifting cultivators is particularly serious in Latin America, where an estimated 50,000 to 100,000 km^2 of forest was destroyed each year during the early 1980s solely for agriculture.[11] About a tenth of all the forest clearance in Amazonia up to 1983 is believed to have been caused directly by small peasant farmers practising shifting cultivation.[12]

Indonesian rainforests are also being devastated by shifting cultivators, who are clearing around 2,000 km^2 each year according to FAO estimates.[13]

3.3b The technique

Shifting cultivation is a traditional practice for forest people.[14] The usual method of clearing land for farming use is to cut down the trees on a plot of high forest and burn them on site. The nutrient-poor soil is then enriched by the ash from the fire, which allows the farmer to establish annual or short-cycle cropping. After a few (generally two or three) years of cropping with good yields, the soil loses its fertility as the nutrient base is depleted, and the farmer abandons the plot. Weeds invade the abandoned plots and secondary succession leads to a natural regeneration of forest cover on the site.

Secondary forest grows quite quickly during this prolonged fallow period. Soil fertility may return after about twelve years' fallow in areas of low population pressure. Meanwhile the farmer continues farming by cutting down, burning and working other plots.

3.3c Sustainability

Under low-population conditions, shifting cultivation can be sustainable[15] and can help rather than hinder natural forest ecological processes. The cyclical pattern, involving forest clearance, short-term cropping, long periods of fallow and forest regrowth, adds further diversity to the naturally rich and varied mosaic of habitats within the rainforests. Traditional shifting cultivation might thus contribute to the huge species diversity of the forests (see section 1.8).

The fallow period is critical in restoring soil fertility (which is paramount for the peasant farmers), but it also plays a vital role in restoring the rainforest ecosystem. The vegetation that grows in the fallow period has a higher biomass and holds a larger and more complete stock of nutrients than the vegetation planted for crops. It also offers much better protection against soil erosion.[16]

In its traditional form, and at realistic levels, shifting cultivation is perfectly sustainable (Figure 3.1a). Native Indians in the Amazon and elsewhere have practised slash-and-burn agriculture for hundreds of years. The key to their success is having a good understanding of the fragile environment they are exploiting. Awareness of natural limits (particularly of soil fertility and yield sustainability) is paramount. Traditional farmers have learned, by trial and error, to keep their operations on a small enough scale to allow the areas left behind as fallow to rejuvenate naturally.

3.3d Increasing pressure

The system works well at low population densities. But as population increases, it becomes necessary to decrease the period of fallow for each plot. Natural restoration of soil fertility and regeneration of forest cover are inhibited. Once the fragile balance between exploitation and restoration which is ingrained into the practices of traditional forest farmers is lost, so too is the basis of sustainability (Figure 3.1b).

Mounting problems from shifting cultivation are being caused by three factors – population increase, immigrant peasant farmers and shrinking forest resources. The natural increase in population in areas dependent on rainforest resources means that there are more people seeking to survive on these rotating allotments. Thus the pressure is on them to grow more during the cropping phase, to reduce the length of

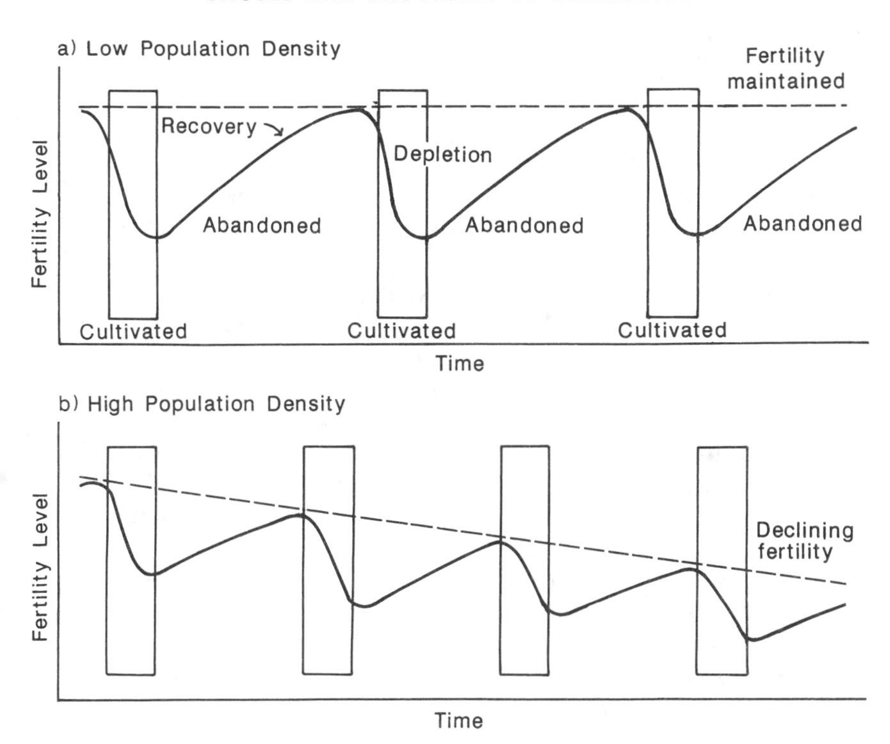

Figure 3.1 Shifting cultivation, population density and soil fertility.

Source: after Goudie (1984)

the fallow phase and to start working new and hitherto uncleared patches of forest.

Much of the extensive forest clearance witnessed in recent years is being carried out by poor landless peasants rather than native forest dwellers. These immigrants are often exiles from overcrowded cities where opportunities for survival if not self-improvement are almost non-existent for them. Few of them are farmers by training or experience, and they lack the traditional wisdom and insight (particularly in sustainable land management) with which native forest peoples are endowed (see section 5.2).

All of this is taking place against a background of shrinking forests to exploit. The overall amount of land available to the growing hordes of shifting cultivators has declined considerably in recent decades. Overworking by peasant farmers is doubtless one reason, but they are more often victims rather than perpetrators of wrongdoings. Governments have often colluded with commercial operators to grant major land concessions within rainforests for logging (see section 3.6) or ranching (see section 3.8). The landless peasant farmers have no formally recognised claims on the forests, so they can only respond by working what forest remains even more intensively.

All three factors conspire to make shifting cultivation a growing problem, particularly in the Amazon rainforest. The problem has been most acute in the Brazilian state of Rondonia. In 1960 the state had a population of 10,000, most of whom were native Indians. By 1985 there were over a million people in Rondonia, many of them recent settlers from inhospitable cities and victims of the country's ambitious but ill-conceived mass resettlement schemes. The population explosion is mirrored in massive clearance of rainforest, which increased thirteen-fold between 1975 and 1985. An estimated one-fifth of Rondonia's rainforests had already been destroyed by the late 1980s.[17] There seems no end in sight to this spiral of clearance as small landless farmers have to keep moving on in search of workable and productive soils, and so new areas of forest have to be cleared just to maintain current population levels

Shifting cultivation must inevitably shoulder some of the burden of responsibility for tropical deforestation, in some areas. But much of the rising pressure comes not from traditional forest peoples but from poor landless peasants who either drift towards the rainforests in the hope of taking control of their own fates and futures, or are transplanted there as a result of government resettlement programmes.

3.4 LAND DISTRIBUTION AND POPULATION PRESSURE

Despite rapid rates of population growth in many tropical countries and the damage caused by collection of fuelwood, most problems arising from over-population stem from land hunger. The two most important factors are unequal land distribution and government resettlement schemes. Both promote the progressive destruction of rainforests, particularly at the margins, as the frontier of shifting cultivators moves relentlessly onwards. Such forces are at work most powerfully in the Brazilian Amazon rainforest.

3.4a Land hunger

Land hunger is created by the confrontation between mounting population pressure and declining land availability. Government policies which promote extensive forest clearance through commercial logging, plantation agriculture and cattle ranching intensify the pressures on remaining forest areas. One result is that forest peoples are pushed out of their natural habitat to occupy marginal areas at much higher population densities. The areas shrink in which tribal peoples can carry out shifting cultivation properly, so fallow time is reduced, soil fertility declines and net productivity in the forest falls dramatically.

But other forces are also at work which intensify the problem of population pressure, particularly in Brazil. Countless landless peasants are driven out of the sprawling cities by poverty. Homesteading and subsistence farming offer these hapless urban exiles the prospect of survival, self-sufficiency and self-determination regarding their own futures (at least in theory). Many other migrants join them, often poor farmers who are lured to the forest with unfulfilled promises of land when they are forced off their own land or are bought out. A common hope they all share is to be able to clear patches of rainforest and scrape a living from the thin soils.

Sometimes these voluntary and involuntary immigrant farmers clear new patches of land in the rainforest. Soil infertility quickly dashes their hopes of being able to farm one area sustainably. They either move on and clear another patch or struggle to survive on their badly depleted soil.

Many migrant farmers take over and rework patches which have been abandoned by earlier logging, cattle ranching and other land uses. From the very start of their activities they face hardship, hard work and much disappointment trying to make a living from already overworked soils. The prospects are not good for them.

The main driving force of mass settlement by peasants in the forest is unequal distribution of land. In Brazil 4.5 per cent of the population controls 81 per cent of the land, and nearly three-quarters of the rural population own no land at all.[18] The forests offer settlers a unique opportunity to claim land for themselves and thus become self-sufficient. They often settle there through lack of alternatives, out of sheer necessity.

3.5 POPULATION RESETTLEMENT PROGRAMMES

Much of the problem of population pressure arising directly as a result of unequal land distribution, particularly in Brazil, reflects passive government attitudes. It also reflects a serious lack of government policies which would seek to solve the problem by tackling the root causes of poverty and redistributing land more fairly.

Active government policies have aggravated the situation even further, by forcing or encouraging migration into shrinking areas of rainforest through resettlement schemes. Many such schemes are ill conceived and doomed to failure from the outset. The settlers are given no education or training in how to cultivate poor soils successfully, so soil erosion and loss of nutrients quickly follow, and land on which so many hopes were pinned is left totally unusable and beyond repair.[19]

The conclusion that many 'colonization schemes have often failed because tropical forest soils are usually too poor to support permanent agricultural settlement'[20] comes as no surprise (section 1.7), and it should have been no surprise to architects of the schemes in Indonesia and Brazil.[21]

Most government resettlement schemes, such as those in Brazil and Indonesia, are designed to provide land and homes for the nation's poor. There are suspicions, however, that such schemes are merely facades for more covert political manoeuvres 'largely intended to secure national sovereignty by establishing a civilian presence in frontier regions',[22] allowing native peoples to be controlled and borders watched.

The two largest resettlement schemes have been in Indonesia and Brazil.

3.5a The Indonesian Transmigration Programme (ITP)

The ITP was developed in 1950. Its main aim[23] was to help reduce the population on the overcrowded islands of Java and Bali by resettling people in densely forested areas on other islands within Indonesia (Figure 3.2). The objective was to move 140 million people over a 35-year period but by the late 1980s only 3.6 million had moved.[24]

It soon became clear that the original programme was both too ambitious and ill conceived. Numerous problems came to light after the first wave of involuntary migrants tried to adapt to their new surroundings. Most settlers were landless city people with little if any farming know-how or experience. They were relocated in extreme and highly unfamiliar environments which posed a severe challenge to them. Few found it possible to work their new land sustainably. Many families were moved on three or more times during resettlement,[25] each move causing yet more rainforest clearance.

Problems were not confined to land management. Settlers were weakened by diseases in this new environment to which they had little natural immunity. Disputes with native forest peoples, who were dispossessed of their tribal lands, were regular and fierce. Other problems they had to contend with included serious food shortages, long droughts and regular rampages by animals like elephants and tigers whose rainforest habitats had been destroyed as part of the programme.

The costs of the failed ITP scheme are difficulty to quantify, but they are high and they are wide-ranging. Vast areas of formerly natural rainforest were cleared or badly damaged through clearance and over 33,000 km^2 of rainforest is now at risk.[26] Tribal peoples have been displaced from their traditional homelands, many of which were re-allocated to the newly arrived settlers. The programme has taken over 2,800 km^2 of tribal homelands in Irian Jaya (Figure 3.2), and forced

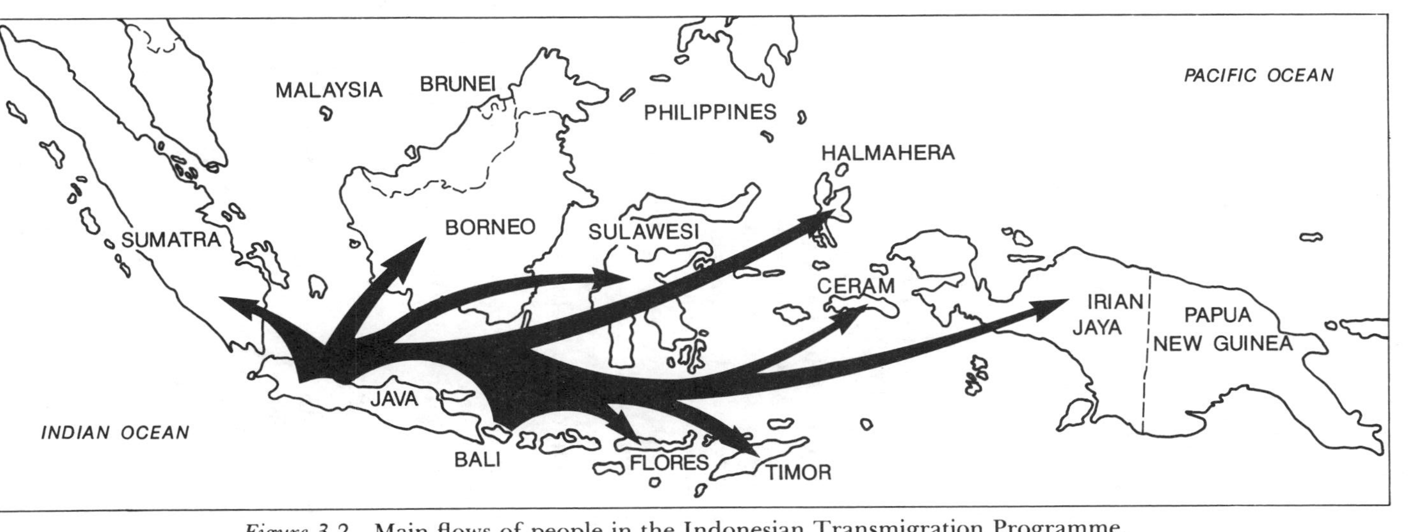

Figure 3.2 Main flows of people in the Indonesian Transmigration Programme.

Source: after Survival International (1989)

10,000 West Papuans to seek refuge from persecution in neighbouring Papua New Guinea.[27]

3.5b Brazilian resettlement programme

Resettlement has been much less structured in Brazil than in Indonesia. There are formal resettlement programmes, but much of the forest clearance in the Amazon arises from informal activities.

Quite often peasants simply claim a patch of forest, clear it and then start farming on it (see section 3.3). It is also common for wealthy landowners to encourage peasants to clear forest off their land in return for the opportunity to farm that land for the first few years after clearance. The land is then handed back to the landowner, who may set up a large ranching or farming operation of his own while the peasant farmer moves on to repeat the cycle somewhere else.

Roads play a very important part in encouraging clearance by settlers, creating corridors of access and strongly influencing resultant patterns of deforestation (see section 3.9). Some roads are built as part of major development projects, while others are left behind by loggers.

Much of the clearance is concentrated along new roads and highways, and settlements spring up where new roads are built into parts of the forest which were previously inaccessible. Settlers gravitate along the new highways and stake their claims. Some may even find a cleared patch on which to start their farming enterprise, perhaps already abandoned by a previous farmer.

The patterns shown in Figure 2.1 are typical, with clearance initially concentrated along major new highways and lateral roads. Through time the frontier of clearance moves progressively further away from the access roads as more and more colonists claim plots. The sad reality is that 'isolation is the most effective protection a rainforest can have, and a road – no matter how primitive – is the end of isolation'.[28]

3.5c Rondonia

One way of attempting to solve some of the problems of severe overcrowding in Brazil's large and rapidly growing cities has been to relocate people. State-sponsored colonisation projects were designed to move peasants from over-populated coastal areas into the Amazon interior. The biggest resettlement project is based in Rondonia, western Brazil (Figure 3.3).

The historical background to the programme provides an interesting example of missed opportunities and inappropriate priorities.[29] In the late 1950s Brazil's President Kubitschek promised rapid economic growth for his seriously underdeveloped country, and saw the conquest

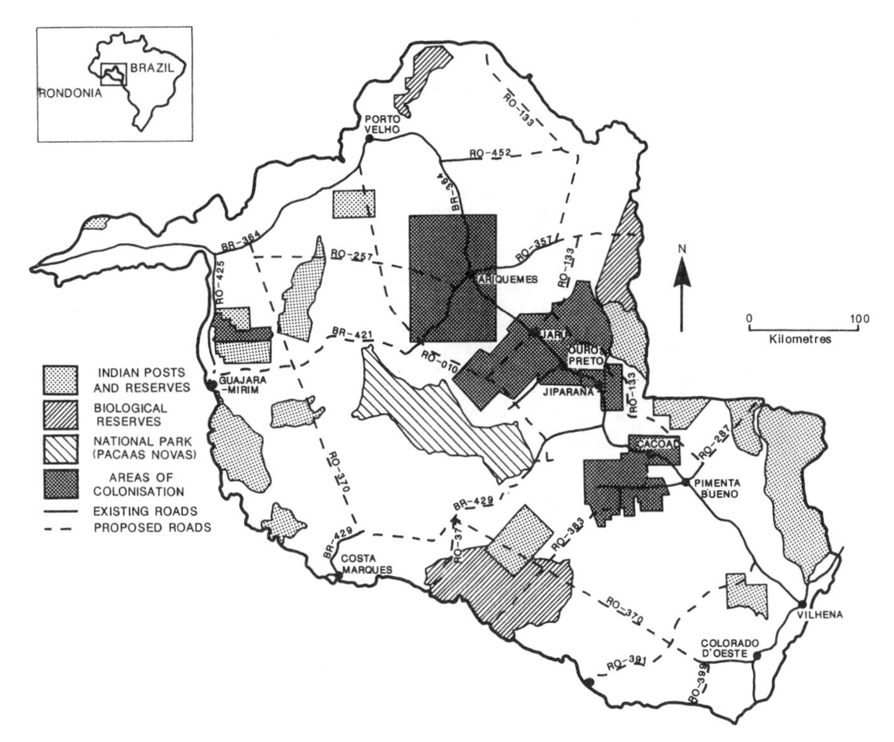

Figure 3.3 The North-West Region Programme, Polonoroeste, Brazil.

Source: after Johnson *et al.* (1989b) and Goodland (1980)

of the Amazon rainforest as an urgent priority. The new capital city of Brasilia was started in 1960 to help this process of opening up Amazonia. A major highway 2,100 km long was built between Brasilia and Belem, requiring the clearance of dense rainforest along its route. In the late 1960s work began on building the east–west Trans-Amazon Highway, which created an access corridor which effectively opened up much of the interior of Amazonia in Brazil.

A major national revitalisation plan was introduced in the mid-1960s, based on encouraging the development of Amazonia by mining, agriculture and cattle ranching. Tax incentives were provided for potential investors by the Superintendency for the Development of Amazonia (SUDAM), and these substantially increased the profitability of investing in major development projects. Investors flocked in, and many landless farmers migrated along the highways in search of land and opportunities. Clearance of rainforest was now taking place with government approval and sponsorship.

Other schemes were introduced to encourage migration of people from the overcrowded and poverty-striken north-east of Brazil, under the widely used slogan 'Amazonia – a land without men for men without land'.[30] The official colonisation agency INCRA made plans to settle a million families along the new Trans-Amazon Highway from Maraba (south of Belem) to Benjamin Constant (on the border with Colombia). Poles of development were established at Maraba, Itaituba and Altamira on the Xingu River. Many incentives were offered, including easy credit, health and education, and technical assistance.

One of the largest development-cum-resettlement initiatives was the North-West Region Programme, Polonoroeste, which began in 1982. The programme, based in Rondonia in north-west Brazil, included the designation of large areas of colonisation and the construction of a series of important new roads to open up the area and improve access (Figure 3.3). It was part-financed with World Bank loans for road building projects totalling US$570 million.[31] The problems were created more by the execution of the project than by its original design, which concentrated on areas with good soils, emphasised perennial crops (like coffee and cacao) and protected some areas for indigenous people and others for biological diversity.

Settlers are provided with 100 ha (1 km^2) plots, accessible from newly constructed base-roads (see, for example, Figure 2.1). The pattern is familiar – nutrient-poor forest soils soon lose their fertility, particularly because few settlers use fertilisers, crop rotation or soil management. Land is soon depleted and further land is cleared.[32]

Between 1970 and 1980 the population of Rondonia grew by 14.6 per cent per year, while the total population of Brazil was rising at a rate of 2.5 per cent per year.[33] The massive influx of relocated urban peoples continued through the 1980s, and between 1984 and 1987 the population of the region doubled to one million.[34]

As people came the rainforest went. An estimated 6,400 km^2 of forest was cleared in the mid-1970s under the Brazilian government's National Integration Programme which settled nearly 13,000 families.[35] Rainforest in Rondonia was disappearing at a rate of about 1,200 km^2 a year in 1975; by 1985 this had risen to 16,000 km^2 a year.[36] Official figures show that 17.6 per cent of Brazilian Amazonia has been lost to state colonisation projects, which far exceeds the 2 per cent of Amazonian soil estimated to be fertile enough to sustain farming.[37]

Like the Indonesian Transmigration Programme, the Brazilian Polonoroeste scheme had little success. Few saw the Amazon as the promised land of opportunity, and less than a twelfth of the planned number of people were resettled. The failure can be related to many factors. Soil fertility is inherently low, and few of the peasant farmers could afford artificial fertilisers. Unsuitable crops were supplied to the settlers, which did not grow well. Little guidance was sought from the traditional shifting cultivators who had successfully grown crops on a small scale for many years. Systems of credit were biased in favour of large farmers, so small subsistence farmers were seriously disadvantaged.

The major outcomes of the ill-fated Brasilian resettlement schemes were a massive increase in the number of subsistence farmers clearing parcels of land in the Amazon, and a corresponding decrease in the area of surviving rainforest.

3.6 COMMERCIAL LOGGING OPERATIONS

Commercial logging poses a serious threat to tropical forests and is responsible for a quarter of the annual loss of primary rainforest around the world. Often it is the primary source of clearance before other land uses take over.

Logging is a major cause of deforestation in many areas,[38] particularly in South-East Asia and Africa. Some intensively logged forests are so badly depleted that tropical timber supplies are running out, with serious economic as well as ecological consequences.

All the major causes of deforestation are ultimately driven by economic motives, but logging differs from the others because it produces high economic returns in the short term only. For example, high quality hardwood, such as mahogany, can be recovered for as little as US$5.50 per m^3 of wood, whereas such wood might fetch up to US$900 per m^3 on the German market.[39]

Export sales of timber bring much more money into developing tropical countries than other commercial uses of the rainforests. But it is not a sustainable type of activity, particularly without massive tree replanting programmes to replenish the depleted timber stocks.

3.6a Rates, areas and volumes

Logging statistics are often expressed in terms of the volume of timber extracted. For example, it is estimated that roughly 3 billion m^3 of tropical forest was felled in 1980 by commercial loggers.[40] The United Nations Environment Programme (UNEP)[41] put the annual production of industrial wood in the tropics at 1.4 billion m^3.

Areas cleared by logging are perhaps more useful to know because this allows comparisons with other causes of deforestation. The figures given in Table 3.2 show that logging accounts for just under a fifth of all rainforest clearance, but more recent studies suggest its share is higher. Between 40,000 and 50,000 km^2 of commercially productive closed forests are logged each year,[42] environmental pressure groups[43] regarding the higher figure as more likely.

Plate 2 Commercial logging in Umbang, Sarawak. Logs are being loaded onto trailers by the Umbang Trading Company. The photograph, taken in 1989, shows clearance of primary rainforest.

Source: Nigel Dickinson/The Environmental Picture Library

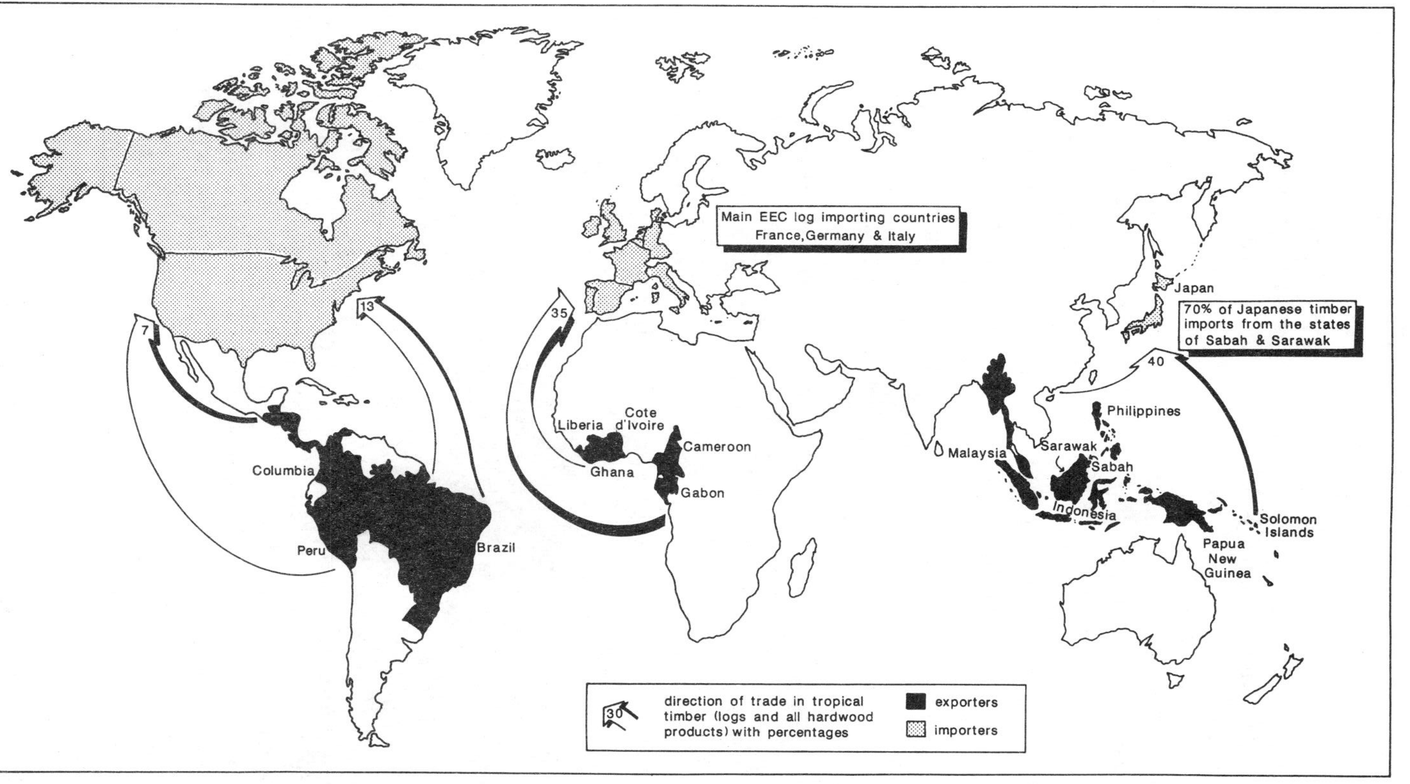

Figure 3.4 Global trade in tropical timber.

Source: after Appleton (1989)

3.6b Problem areas

Not all rainforests are suitable for commercial logging, but where it does occur it can have devastating impacts. Over a third of the rainforest trees in Central America were lost to logging between 1950 and 1983,[44] shrinking the remaining forest from 1.15 million km^2 to 0.71 million km^2. Logging is widely regarded as the second (a distant second) major cause of clearance in the Amazon (after resettlement schemes and encroachment by shifting cultivation). Much of the South and Central American tropical timber is exported to North America (Figure 3.4).

Africa is also badly affected by commercial logging. It provides roughly a third of the world total tropical timber supply, much of which is shipped to Europe (Figure 3.4). Several West African countries (such as the Ivory Coast, Nigeria and large parts of the Congo) had either fully or nearly logged out their forests by the close of the 1980s.[45] The problem in Nigeria is now so bad that timber is imported, whereas previously tropical timber exports brought in much-needed foreign currency.

The worst destruction by logging occurs in South-East Asia,[46] which accounts for nearly half of the world's tropical timber exports (Figure 3.4). Most goes to Japan, which consumes 40 per cent of the world's hardwood supplies.[47] The cost is high. Logging destroys more than four times as much rainforest in Indonesia as organised resettlement or encroachment by farmers.[48] The Philippines were once almost completely covered by rainforest, but by the early 1980s less than a third remained.[49]

Peninsular Malaysia has seen the worst losses. In 1900 it was almost 100 per cent covered by primary tropical forest, which is a valuable capital asset for a country seeking a fast track to development. As the century has unfolded progressively more and more forest has been removed by logging activities. The country now has a massive logging industry. More than half of the rainforest has been cleared by logging since 1960,[50] and some 2,400 km^2 was cleared in 1981 alone. Timber is a major source of foreign income in southern Malaysia, bringing in M$1013.92 million in 1982 and accounting for about a fifth of the state's total earnings.

Logging on this scale clearly cannot continue unchecked. Recent estimates suggest that if deforestation continues at present rates, Malaysia's remaining forest resources will be totally exhausted by the end of the 1990s.[51] Not only is this the end of an irreplaceable ecological resource; it means the end of a mainstay of the country's export economy.

The Malaysian government has tried to reduce the amount of logging by setting logging quotas limiting exports to 41,000 m^3 a year. But the quotas have been largely unsuccessful because they only apply to peninsular Malaysia. Loggers have shifted their attention to eastern Malaysia since the quotas were introduced, bringing a massive increase in deforestation there.

Remaining reserves of tropical hardwoods in Asia are only likely to survive for another forty years if the 1988 removal rates continue.[52] As timber reserves dwindle in South-East Asia the axis of world tropical forestry is likely to shift towards Latin America, where larger rainforests survive and there are fewer government restrictions. But the logging environment is different there. There are different varieties of timber there, it is less accessible, and it is more expensive to transport timber to areas of high demand. It remains to be seen how far the rainforests of Central and South America ultimately fall at the hands (or, rather, the chain-saws) of commercial loggers.

3.6c Logging practices and damage

Loggers normally use one of two methods. The first is clear-cutting (or clear-felling), in which all trees in an area are felled. It is obviously highly damaging to the rainforest. The second method is selective logging, in which only selective trees or species are removed. Given that few tropical tree species are commercially viable, all logging should be selective. If this were the case logging would not be a direct cause of wholesale forest clearance.

In practice selective logging turns out to be almost as damaging as clear-felling, and it destroys vast areas of rainforest. There are a number of reasons. One is that other trees must be cleared to get access to the selected trees, so that even selective logging is accompanied by a great deal of indiscriminate felling. It is estimated that up to a third of the trees are removed just to make way for roads and tracks.[53]

A second important factor is that, unlike temperate woodlands which are more or less single species, the rainforests are highly varied because of their huge species diversity (see section 1.8). This means that commercially valuable trees are widely scattered within the forest, distributed fairly randomly.

There may be up to 2,500 different tree species in a patch of rainforest, but only about 100 will be harvested. Of these, 10 species may account for up to 86 per cent of total production and up to 97 per cent of exported wood.[54] Many of the trees within that patch have no value for sale as timber, but they are still felled. It is argued that logging companies will often slash and burn half a hectare of forest just to get one valuable hardwood tree out.[55]

A third cause of extensive destruction and disturbance, even from selective logging, is the accidental damage to neighbouring trees by felling operations. The damage arises in various ways. Sometimes a tree which is being cut down crashes against adjacent trees and brings them down too. Often the trees topple in groups, because they are effectively laced together by vines and lianas (see section 1.4), so that one felled tree pulls the rest down. Bulldozers are often used to drag felled trees out from the forest onto access tracks and roads. They are sometimes also used to drag or knock trees down. The bulldozers destroy forest soils by compaction and erosion, and they damage other vegetation in the logging area.

There are ways of reducing such damaging activities – such as cutting the lianas and vines before target trees are felled, using helicopters to lift trees out rather than dragging them and using large animals (like elephants or buffaloes) to pull them out. But, driven by the need for high productivity and quick returns on investment, loggers prefer mechanical techniques and rapid operations.

A fourth factor is that even selective logging involves a great deal of wastage of wood, so that much larger areas than might be absolutely necessary are cleared. Studies in Asia have shown that sometimes less than 5 per cent of the cut or damaged timber is actually used.[56] Other studies elsewhere have shown that the loss of 10 per cent of trees by selective logging causes a further 35 per cent of the canopy to be lost, leaving just under half of the original forest intact.[57] Trees left behind might then die of exposure.

The fifth and more indirect way in which logging causes much deforestation is by opening up new areas for other possible forms of clearance. Logging is usually the first operation to push into a new area. Roads required to move the logs can also be used by settlers to enter the previously untouched areas. Opening them of for logging allows homesteaders to claim land through squatters' rights, and then forest patches are cleared for shifting cultivation (see section 3.3). The UN Food and Agriculture Organisation estimates that two-thirds of all primary forest clearance is only possible as result of roads and infrastructure built for logging.[58] Over 90 per cent of the 40,000 to 50,000 km^2 of commercially productive closed forests which are logged each year becomes cropland later on.[59]

3.6d Sustainability

If selective logging really was practised selectively, and only the commercially valuable trees were removed without damaging the others, the forest would be allowed to regenerate naturally by secondary succession after the loggers left. But the large species diversity and the commercial targeting of viable species mean that often large areas of forest are cleared. Apart from being wasteful of the biological capital resource, such practices severely inhibit the ability of the remaining forest to survive and the cleared patches to regrow.

One solution would be for the loggers to clear much smaller patches, and space them out well apart. This would allow more natural reseeding and reduce the areas prone to soil erosion.[60] But, while such a strategy would improve the sustainability of commercial logging in the long term, it conflicts with the loggers' aim of maximising returns on investment in a given area in the short term.

A second approach to enhance longer-term sustainability would involve replanting trees in the areas which have been logged. Ideally, from both ecological and economic points of view, every tree felled would be replaced by a newly planted one. At the start of the logging process companies often plan (or at least say they plan) to replace the trees they remove, but very few do so in practice. Deforestation rates have been up to twenty times higher than replanting rates in recent years.

Whilst selective logging *can* be sustainable, a 1988 survey by the International Tropical Timber Organisation (ITTO) (see section 6.8) found the amount of sustainable logging in practice to be negligible. There are few incentives for commercial logging companies to conserve forest resources, because their logging concessions are short. Most concessions last for ten to fifteen years. Some critics[61] argue that if the concessions were longer – say a minimum of sixty years – companies would have an incentive to protect their own areas of forest from illegal logging and shifting cultivation, and to replant felled areas.

3.6e Market for tropical hardwood

Most of the hardwoods exported by poor tropical countries are used to supply consumer products in rich developed countries. Four-fifths is used in furniture and construction.[62] The list of products is very long, and it includes hardwood veneers on television sets, doors, window frames and furniture. Coffins, chopsticks and crates for Japanese motorcycles are also made from tropical hardwoods.

Conservationists point out that much of the wood is processed into cheap 'throw-away' goods. For example, 80 per cent of the tropical wood Japan imports is made into cheap plywood, which is often used for scaffolding frames and burned afterwards. They also stress that many uses of tropical hardwoods in developed countries are luxury uses, most of which could be substituted by other woods or materials. But fashion and tradition are strong forces contributing to inertia.

So long as international demand for tropical timber remains high, logging will continue to claim new areas of tropical rainforest.

3.6f Economics and ecology

The damaging commercial logging operations are driven by market forces and feed the large and growing international market for tropical hardwoods. Demand for hardwood has risen sharply in developed countries since about 1945, so they now consume much more tropical timber than the developing countries. Hardwood consumption in the developed world has risen by fifteen times since 1950, compared to a threefold increase in demand in producer regions.[63]

Some argue, therefore, that the ultimate responsibility for this pressure on the rainforests rests squarely with the developed countries for creating a demand for tropical hardwoods which is entirely incompatible with the ability of tropical countries to supply them in an ecologically sound and sustainable way.

The biggest consumer of tropical timber is Japan, which imports 15 million tonnes of wood and nearly three-quarters of all processed wooden goods each year.[64] The irony is that Japan could quite readily supply all of its timber needs from domestic sources, but it chooses instead to conserve them and import three-quarters of its timber from the tropical forests of South-East Asia.[65] Imported tropical wood is cheaper and is regarded as expendable, while Japan grants protected status to its own trees.

But part of the equation which continues to promote logging of tropical rainforests is that the producer countries rely heavily on timber exports to generate foreign revenue which helps them to pay off massive international debts (often for major development projects; see section 3.9). They are locked into a development system which they can only service by selling their biological capital assets – the rainforests.

There is no doubt that 'in the short term, it is more lucrative to cut down the forests than to conserve them. Marketing sustainable rainforest products such as fruits, genetic material and rubber gives a lower immediate return than destroying the forest for its timber.'[66] Sales of tropical hardwood world-wide in 1989 totalled US$7 billion, three-quarters of which was earned by Indonesia and Malaysia.[67]

Much of the tropical timber is exported as unprocessed logs. But governments are starting to appreciate that they can make better profits by processing the timber before it is exported. The export of unprocessed logs has been stopped or dramatically reduced in the Philippines, Malaysia and Indonesia. New policies favouring the establishment of value-added industries (craft industry and manufacture of wooden products such as finished window frames) are creating employment and helping local and national economies.

Logging pressures are being further aggravated by large projects, often supported by international financial aid, which promote clear-cutting of tropical forests for the woodchip industry. Between 1982 and

1987, for example, Indonesia planned five major pulp projects to supply both internal and external markets.[68] Commercial returns from such ventures are likely to be low, but ecological and environmental costs are guaranteed to be high.

But the same economic forces which promote logging of the rainforests also promote over-exploitation. Short-term profit seems to far outweigh long-term sustainability in the minds of the commercial logging companies and until recently also in the minds of many tropical governments.

Attitudes are changing as timber supplies in many tropical countries are being exhausted. The economic (and hence political) impacts of supply exhaustion are significant. Exports from the main tropical hardwood producers peaked at about US$6.8 billion in 1980 and fell to $5.8 by 1985. Exports are expected fall further, to under $3 billion by 1998, as 23 of the 33 main exporting countries exhaust their forest resources.[69] The World Bank predicts that only 10 of these 33 countries will still be able to export tropical timber in the early 2000s.

3.7 PLANTATIONS AND CASH-CROPPING

Many patches of rainforest are being cleared to create land for cash crops and plantations. As with commercial logging, the economic objective is to maximise returns on investment in this unproductive environment. Most of the products are exported to foreign markets, too, and thereby earn valuable foreign exchange.

This is a much less homogeneous pressure on the rainforests than logging because the patterns, scales and products vary a great deal from place to place.

3.7a Cash-cropping

Many schemes involve cash-cropping. In Thailand large areas of rainforest have been cleared to create land for growing cassava (manioc), the source of tapioca but also a calorie-rich feed for cattle, pigs and poultry in the European Community.[70] Elsewhere cash crops such as pepper and cocoa plants (for making chocolate) are grown as monocultures for export.

Brazil is a major food exporter and forest clearance for crops continues. Yet Brazil also imports vast amounts of food. It seems that wealthy farm owners prefer to grow exotic crops for sale abroad at high profit rather than grow basic food crops for sale at home.[71]

Another threat to the Amazon comes from the increasing (but illegal) cultivation of coca, particularly in Peru. The leaves of the coca plant are used to make cocaine, a hugely lucrative product on the international black-market. Farmers are prepared to take the risks involved in burning patches in virgin forest on which to plant and grow coca, in clandestine operations. Latin American governments (with support from the United States) have tried to eradicate the crops by spraying them with herbicides, but spray drift often damages the surrounding rainforest. Cultivation of coca is believed to account for the clearance of 6,800 km^2 of Peruvian rainforest this century, roughly a tenth of the total deforestation in Peru.

3.7b Plantations

Other forest clearance schemes create land for plantations. Rubber and oil palm plantations have replaced natural rainforests on lower hill slopes in many parts of Malaysia, the produce destined for use in developed countries.

Parts of Amazonia are also being replaced by plantations. Under the Grande Carajas project (see section 3.9), for example, some 55,000 km^2 of rainforest has been cleared for export-orientated plantations and biomass fuel farms. Government-backed plantations are also being set up in Rondonia.[72]

3.7c Mono-specific farming

Cash-cropping and plantations almost invariably involve single species stands. Experience in most environments shows that mono-specific farming is usually beset with problems. Some relate to the susceptibility of large mono-specific farms to disease and pest attacks, which can quickly kill all plants. Other problems stem from market uncertainties, particularly if over-production of a cash crop leads to market saturation and price falls.

In the unforgiving tropical environment, with poor soils, high temperatures and abundant disease and pests, the problems of monoculture are amplified. Little wonder, therefore, that most initiatives fail.

History is full of examples of failed attempts to raise sustainable plantations in the tropics.[73] Early this century, for example, Henry Ford invested US$20 million in unsuccessful attempts to establish rubber plantations in Amazonia. A key factor was the poor rate of tree growth, caused partly by infertile soils and partly by South American leaf blight. Poor soils and pest attacks also led to the failure of an ambitious attempt by American entrepreneur Daniel K. Ludwig to establish a 10,000 km^2 plantation of fast-growing trees for pulp and high-yielding rice strains along the floodplain of the Jari River. Ludwig bought the land for $3 million in 1967, launched a timber, woodpulp and agricultural project, but eventually lost nearly $1 billion on the failed venture.[74]

3.8 CATTLE RANCHING

Cattle ranching as a major cause of rainforest destruction is more or less confined to Latin America, where cattle grazing has become a major land use. Latin America saw a threefold increase in the area of pasture and ranchlands between 1950 and 1980.[75] Clearance for cattle ranching continues at a rate of around 20,000 km^2 a year.[76] If recent trends continue there will be no rainforest left in the region by the year 2000.

Costa Rica is one of the worst affected Central American states; two-thirds of the rainforest destruction there has been attributed to cattle ranching.[77] But it is in Amazonia that the problem has been most acute. Brazil has had a pressing need for export earnings since the 1973 oil crisis and price rises, and vast amounts of private investment and many major development projects centred on Amazonia were approved by the government. Big cattle ranching projects were an appealing way of securing overseas investment, and new road and highway developments encouraged the diffusion of deforestation for ranches from the north-east of Brazil into the heart of Amazonia. Nearly three-quarters of all forest clearance in Amazonia up to 1980 has been attributed to cattle ranching.[78] Relatively little of the Amazonian beef is exported.

3.8a Reasons for growth

Cattle ranching has grown quickly as a major threat to the rainforests, particularly in the Amazon. It is likely to continue growing in the foreseeable future. A number of factors explain the growth.

Cheap land and government subsidies

Ranching is a lucrative business only when economic viability in Amazonia has been further enhanced by cheap land deals and government subsidies. Large companies have been given land at subsidised prices by the Brazilian government to encourage them to invest in cattle ranching there. Rainforest along the route of the Trans-Amazon Highway was opened up to large-scale entrepreneurs (mainly cattle ranchers) in 1973. Plots of between 5 km^2 and 550 km^2 were made available further back from the road than the 1 km^2 (100 ha) plots for peasant farmers. In 1981 these larger plots occupied three and a half times the area of the small farm plots,[79] a total of 27,000 km^2.

Among the major investors in Amazonian cattle ranching have been large multinational corporations like Volkswagen (1,390 km^2 in 1983) and Armour-Swift (720 km^2 in 1983).[80] Multinationals buy up large tracts of rainforest land in developing countries at very low prices compared with land costs in developed countries. The land is often logged first, with logging revenues going directly to the multinational company, not the developing country. After logging has cleared the forest, the land is then used by the company for the purpose for which it was bought (usually cattle ranching).

Part of the key to understanding why major corporations invest in cattle ranching is the rapid rate of appreciation of land values in Amazonia. Through the 1960s and 1970s, land in Amazonia more than doubled in value each year.[81] Under such conditions ranching is a very good investment, particularly because cattle require a relatively small input of capital. Ranchers can often afford to operate at a loss on the cattle business while they sit on the growing land asset.

Government tax relief and tax holidays

Tax exemptions for ranching schemes have made encroachment into rainforest even more commercially attractive. Through the late 1970s the Brazilian government offered large financial incentives to establish ranches in Amazonia. In 1977, for example, 72 per cent of all investments in cattle in Brazil were in the form of incentives, loans and subsidies. The incentive schemes were discontinued for new ranching enterprises established after 1979, but remained for projects previously approved.[82] Between 1975 and 1986 over £1 billion was paid out in subsidies to ranchers in Amazonia.[83]

The main subsidies from which cattle ranchers in Amazonia benefited were tax credits and tax holidays.[84] Tax credits were highly attractive. Corporations were allowed exemption on half of their federal tax dues so long as they invested an equivalent amount in Amazonian development projects under the Fundo de Investimento da Amazona (FINAM). Such credits could be worth up to 75 per cent of the project cost, and are estimated to have totalled US $1.4 billion between 1965 and 1983. Some 42 per cent of the credits went to 470 cattle ranches.[85] Tax holidays are also very effective. Those who invest in the FINAM projects are eligible for a fifteen year tax holiday (that is, a tax-free period) on income derived from expansion or modernisation.

With such attractive financial incentives, it is little wonder that by the late 1980s large cattle ranches were occupying 72 per cent of the cleared land in Amazonia.[86]

Demand for beef in the USA

A major catalyst for rainforest clearance to create cattle ranches has been the large, buoyant and lucrative foreign market for the beef, especially in the USA. Most of the beef (apart from that produced in Amazonia) is exported; Latin American beef is roughly half the price of US beef.[87]

Up to 90 per cent of the exports from Latin America end up in North America, where much is made into hamburgers for fast-food chains. This is the so-called 'hamburger connection'[88] which conservationists want to see stopped.

Domestic consumption of beef in most Latin American countries is negligible compared with exports to the USA. Beef consumption in Costa Rica fell by over 40 per cent between 1960 and 1979, while production increased threefold over the same period.[89] A similar pattern occurs throughout the whole of Central America, where beef consumption has fallen while production has been rising.[90]

Although the 'hamburger connection' may apply to Central America, it does not apply to the Amazon.[91] Beef from the Amazon is not safe to eat and is prohibited from US markets. Ironically, Amazonia is a net importer of beef. Cattle ranching is important there for other reasons, more related to land speculation and tax evasion than to cows.

3.8b Cattle ranching and forest clearance

Cattle ranching is a very land-hungry operation which requires the clearance of large areas of rainforest. Yet the inherently poor soils (see section 1.7) mean that cattle grazing is neither a suitable nor a sustainable use of converted forest land. Soil fertility and yield fall quickly after ranching starts, and even a few years of intensive cattle grazing can destroy the physical and chemical structure of the soils. Around 15,000 km^2 of grass pasture had been planted in the Amazon region by 1977, yet by 1978 a third of the pasture was badly degraded or had been invaded by secondary growth of vegetation.[92]

Rainforest soils are not suitable for animal grazing or crop production and they quickly deteriorate. The number of cattle they can sustain falls sharply over a ten-year cycle. Ranches are often run for between 5 and 8 years, then the rancher moves on and starts the process again elsewhere.

Opening up more land is often seen as cheaper than trying to prolong commercially viable use of existing ranchland by adding substantial quantities of fertilisers and investing in techniques for soil conservation, disease control and improved breeding.[93] Land is cheap so ranches are often abandoned and re-established elsewhere in virgin cleared forest lands.

There are lots of similarities with shifting cultivation (see section 3.3), but grazing is much more damaging. Soils are often leached beyond repair, preventing any subsequent natural regeneration of forest.

One of the most serious problems associated with cattle ranching is the economic loss to the host nation. The developing country makes no money from the transaction other than the price the land was originally sold for; all profits from ranching are enjoyed by the owners of the ranches.

Ranching continues to grow in Latin America and is fast becoming more widespread in Asia and Oceania. The trend seems difficult to reverse, especially given low land values, governments willing to offer financial incentives to promote investment in ranching, and the ever-ready export market for beef.

3.9 LARGE-SCALE DEVELOPMENT PROJECTS

Amazonia, and to a lesser extent other major rainforests around the world, is also being cleared with government approval through large-scale development and construction projects.[94] These are designed to provide the basic infrastructure needed for the developing country to improve itself and enhance its international standing. The two most important types of project, particularly in Amazonia, are road building schemes and dams to provide hydroelectric power. Both cause massive and widespread forest clearance.

3.9a Road building

Road construction causes damage to the rainforest both directly, through the construction process, and indirectly by opening up new areas which are then subject to other clearance pressures.

The most ambitious road building project through a large rainforest is Brazil's Trans-Amazon Highway (Transamazonica on Figure 3.5). The 3,300 km highway was announced in 1970, when the Brazilian government decided to integrate the Amazon with the rest of Brazil through a road network.[95] There were three aims for the highway.[96] It was to act as a safety valve for the poor and overcrowded north-eastern area of Brazil, by attracting peasants from that region. Secondly, it was to open up the Amazon, which had low population (only 4 per cent of Brazil's population in half its area, in 1970). Thirdly, it was to allow access to minerals and timber which would fuel Brazil's planned 10 per cent annual economic growth.

Financial and material assistance for the scheme came mainly from the United States.[97] Loans from the Inter-American Development Bank and the World Bank financed the project, USAID provided grants-in-aid for technical help and loans, and the US Army provided material help in building the road.

The highway project failed to achieve its resettlement objective because many fewer peasants than expected left the north-east to settle in Amazonia. The project was also a financial disaster which cost vastly more than expected. The forecast cost was calculated at a time when oil

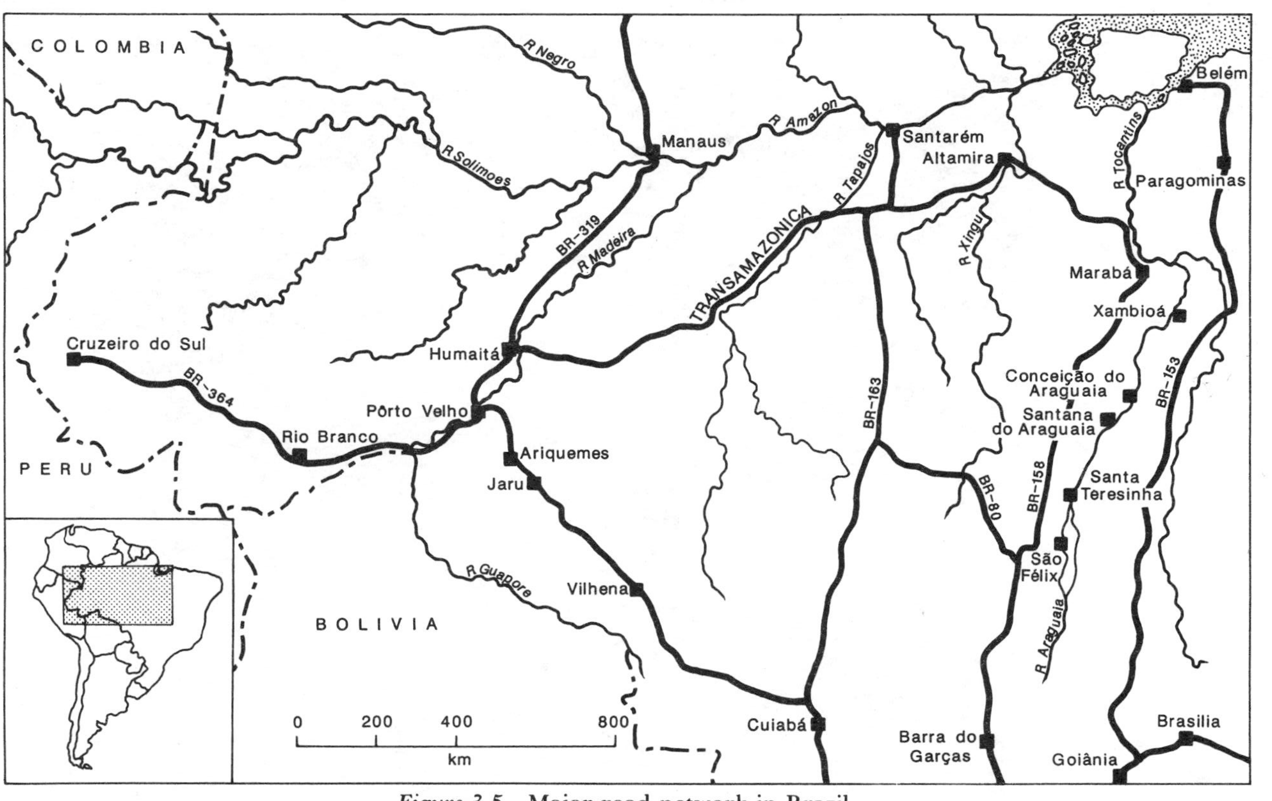

Figure 3.5 Major road network in Brazil.

Source: after Branford and Glock (1985)

Plate 3 Road building in the Brazilian Amazon. A strip of primary rainforest is being cleared to create a corridor for road building. The direct damage is relatively small, but increased access will encourage shifting cultivators and other agents of clearance into the area, who in turn are likely to clear much bigger areas of forest.

Source: Susan Cunningham/The Environmental Picture Library

was $2 a barrel, but by 1981 Brazil was already paying over $35 a barrel for its oil. By 1981 the cost of the highway had escalated to nearly $500 million.[98]

A central component of the Polonoroeste resettlement programme in Brazil, during the early 1980s, was the westward extension of the Trans-Amazon Highway.[99] The extension was built between Cuiaba (in Mato Grosso) and Porto Velho (in Rondonia) (Figure 3.5) and it encouraged a flood of migrants into the Amazon in north-west Brazil. During the early 1980s population was increasing in Rondonia at a rate of 16 per cent a year. The rainforest suffered directly, and between 1982 and 1985 deforestation rose from 5 per cent to 12 per cent.

The situation in Rondonia quickly became serious. An estimated 200,000 peasant farmers were locked in an endless cycle of forest destruction as newly cleared land failed after two to three seasons under cultivation. The wisdom of Brazil's resettlement scheme was questioned, opposition to the forest clearance grew rapidly, and pressure was mounting to halt the destruction. After particularly vociferous protests in 1985, the Brazilian government and the World Bank (which had sponsored much of the road building programme) decided to cease these massive road building and road improvement schemes. Attention was switched away from development towards conservation, as the Brazilian authorities worked on plans to protect wildlife and native Indian reserves.

The 1985 pause in developing a major highway system through Amazonia was just temporary, given the strong and growing need for Brazil to develop its infrastructure to help it to start paying off massive loans and debts to international banks and agencies. In June 1987 the Brazilian and Peruvian Presidents met to approve further extension of the Trans-Amazon Highway into Pucallpa in the Andes, which would link it to Peru's road network. The plan was for the highway eventually to stretch westwards all the way to the Pacific, financed by loans and grants from Japan.[100] By late 1991 this scheme had been shelved, but revival is possible in the future.

3.9b Minerals and energy resources

Further threats to the Amazon rainforest come from mining of the rich and varied store of minerals locked in its basement rocks and surface deposits. Brazil earns some $9,000 million a year from mining of minerals such as gold, diamonds, uranium, titanium and tin, much of it from Amazonia.[101]

Some of the worst environmental damage has been associated with a recent gold rush in the north-west Amazon region of Rio Negro (see section 5.5). The Brazilian government has granted mining concessions

Figure 3.6 Location of the major development projects in Brazil

Source: after Friends of the Earth (1989c) and Fearnside (1989)

on 2,000 km^2 of rainforest, but damage is being caused over a much wider area by the speculative and unregulated activities of an estimated half a million prospectors scouring the Amazon for gold. Miners use mercury to extract gold from the river muds, and this had led to the poisoning of communities downstream.

The rainforest is also threatened by the Grande Carajas project in north-east Brazil and by the recent discovery of oil and natural gas reserves in the interior of Amazonia (Figure 3.6).

3.9c Dams and hydro-electricity schemes

Major dam construction projects designed to exploit the hydroelectric power (HEP) potential of steep and well-watered rainforest rivers pose a serious threat to the rainforests. Like road building projects the HEP schemes affect forests in a variety of ways, including construction of the dams and access roads, flooding of forests as reservoirs behind the dams and opening up the forest to other types of development and encroachment.

One recent example which received widespread publicity round the world was the proposal to clear a large area of Tasmanian rainforest around Lake Pedder to build the Franklin Dam.[102] The scheme was subsequently stopped after a series of public protests.

Brazil has launched a major programme to build a series of HEP dams along tributaries of the Amazon.[103] These steep, fast flowing rivers have the potential to generate 100,000 megawatts of power. The Brazilian government aims to use the dams to produce cheap electricity as an incentive to major foreign investors, who will receive tax concessions and be able to buy electricity at 30 per cent below the market price.

3.9d Grande Carajas project

Showpiece of the programme is the massive development project at Grande Carajas (Figure 3.7), which includes hydro-electricity dams, mines, metal processing, forestry, ranching and farming.[104] The programme envisages building dams capable of supplying up to 40 per cent of Brazil's electricity (22,000 megawatts) by the year 2000, which would be a major asset to the Brazilian economy.

The Grande Carajas project is based around a major iron smelting complex at Carajas, backed by a $600 million loan from the European Community. The pig iron smelting furnaces there use iron ore from a local open-cast mine (Figure 3.7). Carajas has one of the world's largest mineral deposits, with known reserves worth US$500 billion. The deposits include 18 billion tonnes of high grade iron ore, as well as gold, bauxite and other minerals. A new 800 km railway has been built to a new port at Sao Luis. Only 100 km^2 of forest had to be cleared for the mine itself, but plans include 100,000 km^2 of farming and forestry projects.[105]

The Carajas project itself has been managed very carefully and has caused little direct forest clearance. But the associated activities in the area have produced numerous problems. Even worse are the unrelated Grande Carajas hydroelectric developments.

There is also a bauxite mine which produces 8 million tonnes per annum, and an aluminium smelter with an export capacity of 800,000 tonnes of aluminium and 20,000 tonnes of aluminium oxide (mostly destined for Japan) each year. More than 6,000 km^2 of rainforest must be cleared each year to provide enough charcoal for the smelters.[106]

Major dam schemes have been completed at Tucurui and Balbina, which have flooded a total area of roughly 4,500 km^2 of rainforest in Amazonia.[107]

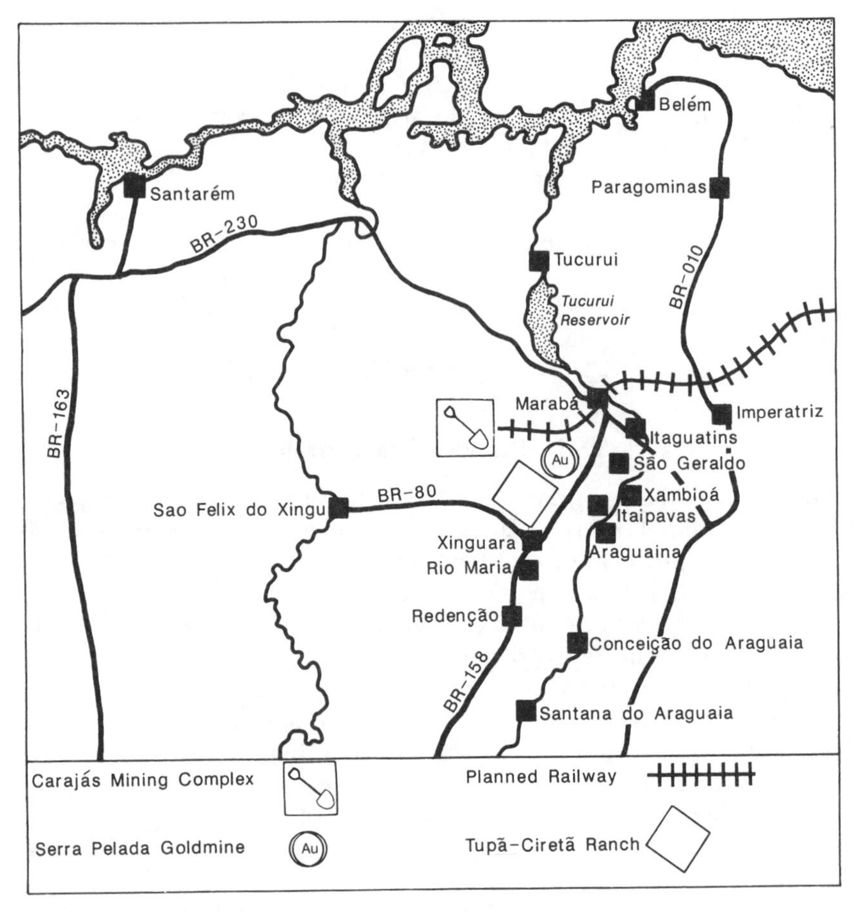

Figure 3.7 The Grande Carajas project, Brazil.

Source: after Branford and Glöck (1985)

It is estimated that the Tucurui HEP project alone could cut by half the country's $9 billion annual oil import bill.[108] The dam at Tucurui (Figure 3.7) is 19 km long and when completed in 1984 it was the fourth largest dam in the world. Around 1,800 km^2 of virgin rainforest was flooded but it was not cleared. Foliage was killed by spraying it with dioxin and the trees have since rotted, clogging the HEP turbines.[109] The Brazilian authorities have since realised what a waste of biological assets this approach (designed to reduce costs by speeding the filling operation) was. An estimated 2.5 million m^3 of prime timber, much of it commercially valuable, was submerged in the flooding.[110] Millions of animals were drowned as the reservoir filled. Large numbers of indigenous people, the Parakana Indians, lost their land and were resettled elsewhere.

The Balbina HEP scheme near Manaus in western Amazonia (Figure 3.6) flooded nearly 2,350 km^2 of rainforest. The dam closed in October 1987, and like Tucurui the trees were not removed before the reservoir was filled.[111] Balbina is only a few metres deep. An estimated 6.8 million m^3 of timber was flooded because the timetable allowed for logging was short (two years) and the wood volume was regarded as too low to interest potential logging contractors. Balbina also led to the displacement of up to a third of the surviving members of the Indian tribe the Waimiri-Atroari.

Like many development projects in developing countries, the cost of Balbina escalated through time. The scheme cost $750 million, nearly twice the original estimate ($383 million). All the money was borrowed on the international market. The overall effectiveness of Balbina is open to debate, too, because it provides only one-third of the power supply to Manaus. This proportion will decrease as the city's population continues to expand.[112]

3.9e Altamira complex

Another vast HEP project is scheduled for construction within the Xingu River Basin at Altamira, west of Grande Carajas (Figure 3.6). Project costs are estimated at $10.6 billion, nearly a tenth of Brazil's foreign debt.[113] The Xingu River Basin HEP project plans to produce 17,000 megawatts (MW) of electricity from five dams flooding 18,000 km^2 of rainforest. A third of the electricity would be used within the Amazon Basin; the rest would be carried vast distances by transmission line to south-east and north-east Brazil.

The centrepiece of the Altamira project would be two large dams. Babaquara would flood 5,600 km^2, to become the world's largest artificial lake, and Kararao would flood 1,225 km^2. An estimated 70,000 people would be displaced. The World Bank, a major sponsor, has revealed that the original plans for Babaquara have been scrapped because of its likely environmental and social impacts, as well as its economics. The Bank also argues that Kararao would not be viable without Babaquara.[114]

3.9f Problems

Apart from the vast economic investments in these HEP schemes, which appear to be millstones around the neck of the Brazilian government, the dams have faced some serious environmental problems. Clearance of forest from the reservoir area before flooding is generally viewed as uneconomic for the companies building the dams, but this simply defers costs from construction to subsequent operation and repair phases. Many of the dams are silting up faster than expected, which is reducing their life-expectancy. Fish life in rivers downstream from the dams is often seriously disrupted, along with other species in the forest ecosystems surrounding the reservoirs. The fluctuating water levels in the reservoirs provide ideal breeding grounds for mosquitos and other hosts that carry debilitating diseases like malaria and schistosomiasis, so that human health can be adversely affected over a wide area.

3.9g Energy plans

The HEP dam projects are key components in Brazil's long-term energy planning. The ambitious Plano 2010 (2010 Plan) requires 136 new HEP dams, 68 of them in the Amazon. The plan aims to provide a third of Brazil's energy from the Amazon by 2010.[115] It would consume tens of thousands of km^2 of virgin rainforest (as much as 250,000 km^2 according to some estimates),[116] displace hundreds of thousands of people (up to half a million according to some estimates) and require massive international funding.[117] The subsequent 2020 Plan was designed to supersede the 2010 Plan, and would mean the construction of 80 dams in Amazonia.[118]

Since the late 1980s there has been broad based international debate about the benefits and costs of such vast development schemes which consume so much rainforest. Environmentally destructive projects funded by the World Bank which have been debated[119] include Indonesia's Transmigration Programme (see section 3.5), Power Sector loans to Brazil (see section 3.10) and the Brazilian Grande Carajas project (see section 3.9d).

Key actors in the debate have been the World Bank (provider of many of Brazil's loans) and the US Environmental Defense Fund (defender of the rainforest and the interests of rainforest peoples). Other international environmental groups such as Friends of the Earth, Greenpeace, the World Wide Fund for Nature and the Rainforest Foundation have supported the forest defenders.

Much of the debate has centred on accusations that the Brazilian agency FUNAI (the National Indian Foundation) has neglected its responsibilities to protect the forests and their peoples, and that the World Bank has persistently failed to enforce the environmental promises made in respect of its loans. By the close of the 1980s the World Bank had changed its policies (see section 3.10) and was not granting new loans because of Brazil's failure to protect the forest and its native Indian peoples.

3.10 FOREIGN AID

A common thread through much of this chapter has been the tension within developing countries, particularly Brazil, between protecting their rainforests and capitalising on them as a valuable resource. The rainforest problem is not confined to the tropical countries themselves, because of the many ways in which developed countries are implicated. These include consumer-led market forces (for tropical hardwoods, for example; see section 6.7) and the search by multinational corporations for high-yielding investment opportunities (through cattle ranching, for example; see section 3.8).

At the same time the developing countries borrow heavily from banks and agencies in the developed world to finance major development projects. This creates serious debt crises for them, because they are often unable to maintain high interest payments let alone start to repay capital sums. There is every temptation for them, therefore, to 'mine' their forests as capital assets to pay off foreign debts.

Brazil's experiences are particularly revealing.[120] The ambitious 2010 Plan requires vast international funding. Between 1980 and 1988 Brazil received over $2 billion in loans from the World Bank and the Inter-American Development Bank. The first loan (Power Sector I) totalled $500 million, some of which went to fund previously abandoned schemes such as the massive Tucurui Dam (which had been refused funding by the World Bank in the late 1970s as ill conceived).

The proposed second loan (again of $500 million) (Power Sector II) was scheduled for approval by the World Bank in mid-1987, but it was repeatedly delayed. Two reasons were given for the delays – poor economic efficiency in the first round of projects, and failure to comply with the environmental and social conditions which were attached to the first loan.[121]

A further loan of $600 million was agreed in September 1988 by a consortium of European and North American banks as part of a $5.2 billion debt rescheduling package. This new loan was conditional on the approval and use of the World Bank's second loan. Like the first loan, this 1988 loan was not earmarked to cover particular projects. Observers expected the Brazilian government to use at least part of the 1988 loan to complete or start construction of HEP dams (including the Altamira complex; see section 3.9e).

3.11 CONCLUSIONS

A number of clear conclusions emerge from this review of the main causes of rainforest clearance. Many relate to the problems of data availability. It is very difficult, for example, to establish exactly how big the problem of clearance actually is. The evidence on how much rainforest is being cleared, and where, is rather patchy even today (although the information base is improving fast).

Neither is it easy to say categorically why the forest is being cleared, because the reasons vary from one area to another. The pressures on the forests also change through time, making it difficult to draw meaningful comparisons between studies of different ages.

Apportioning blame between the different sources of pressure on the rainforests has its problems, too. Multiple causality seems to be much more important than some studies suggest, because some forms of clearance are encouraged by others. Road construction opens up previously inaccessible portions of forest to loggers, for example. Shifting cultivators often take over land abandoned by commercial loggers or by cattle ranchers. Dam construction projects require road building schemes, as well as flooding large areas of rainforest directly.

The complexity reveals much competition for land and space within the forests, so it is difficult if not impossible to single out any particular group(s) and lay the major blame on their shoulders.

Another way of looking at the problem is to ask, who benefits from the clearance of the rainforest? This is a bit easier to answer, because it is never the native tribal peoples and rarely is it the subsistence-level peasant farmers who engage in shifting cultivation. Even national governments gain much less than they would like to from clearance of their own forests. There are two main groups of beneficiaries. One is the small but wealthy group of domestic elites, who have access to political support and invest in major development projects (such as cattle ranches). The major beneficiaries are usually foreign. Foreign consumers gain from access to relatively cheap resources (like tropical beef and hardwood). But the best benefits are enjoyed by the multinational corporations which invest in rainforest development schemes – the commercial loggers, the international cattle ranchers and sponsors of major dam construction projects.

The two root causes of tropical deforestation are the pursuit of short-term gain by international investors, and the quest for development status by the developing countries. Paradoxically, their activities in encouraging clearance of the rainforests are not sustainable in environmental terms, and do little to help ease the plight of the millions of poor, landless people within their territories.

In this chapter we have seen examples of the impacts of forest clearance caused by different types of activity. These illustrate some of the local and national costs of deforestation. But there are also broader impacts which are more significant on the regional and global scales. We turn to examine these next, in Chapter 4.

4

IMPACTS AND COSTS OF DESTRUCTION

They took all the trees and put them in a tree museum,
Charged all the people a dollar and a half just to see them.
Don't it always seem the same,
you don't know what you've got till it's gone . . .

Joni Mitchell, *Big Yellow Taxi*

4.1 INTRODUCTION

Rainforests represent one of the world's most valuable non-renewable resources,[1] and to throw them away is inexcusable. Yet the rapid pace of clearance over the last ten years, coupled with the many places where clearance is taking place (Chapter 2), paints a bleak picture about the future prospects of the remaining rainforests. Deforestation is one of the most important environmental problems in developing countries.[2]

When rainforests are cleared, for whatever purpose, it is much more than just a stand of trees or a pile of wood that is lost. The losses are not confined to the developing countries where rainforests grow, either. In this chapter we examine the broader impacts of deforestation, particularly those with an international or even a global dimension.

4.1a Losses and losers

Some of what is lost when rainforests are cut down or burned reflects their intrinsic value (see section 1.9). Most of us are never likely to be able to set foot in a real rainforest, but that is not the issue. What matters is that the rainforest is important quite simply because it is there. Whether or not it has some inalienable right to survive is a debating point in moral philosophy; the fact that we would miss it if it disappeared completely is a matter of human nature and sensitivity.

The extrinsic values of rainforest (see section 1.9) are easier to quantify and thus they form the basis of any evaluation of the impacts of deforestation. Whatever utilitarian interest we might have in nature's richest ecological storehouse (including many interests we were probably totally unaware of, such as using drugs and medications derived directly from rainforest products), we are all losers as the remaining forest reserves dwindle and disappear.

Inevitably the most direct losers are the people for whom the rainforest is a home. Traditionally an estimated 50 million people live in the world's rainforests. These are mainly tribal peoples whose lifestyles and cultures are tightly interwoven with the natural cycles and processes of the forest, and who have adapted over many generations to life in the forest. We consider their plight in Chapter 5.

Here we concentrate on the main impacts of deforestation, which include the loss of biodiversity and natural resources, loss of environmental services and possible changes to climate on the local, regional and global scales.

4.2 LOSS OF BIODIVERSITY

Biodiversity is a shorthand expression for biological diversity,[3] and the loss of biodiversity arising from deforestation is caused by the extinction of species.

Natural rainforests have incomparably rich and varied plant and animal populations which make them the world's most diverse ecosystems (see section 1.8). But underlying that richness is a fragility and susceptibility to change. The loss of biodiversity has two main consequences – the loss of natural species (an ecological loss), and the loss of forest resources (a utilitarian loss).

4.2a Species extinction

The biggest problem associated with clearance of the world's rainforests is species extinction. Whilst the pace of extinction has doubtless speeded up in recent years, the problem itself is not new. The danger of mass extinction of rainforest species if intensive use and clearance continue was pointed out over two decades ago.[4] Yet clearance continues, and at an ever-increasing rate (see Chapter 2).

Extinction occurs when the last surviving individuals within a species die. Before that terminal point is reached, however, population levels decline as progressively more individuals die and rates of reproduction fall.[5] Detailed biological monitoring can sometimes detect these early warning signs. Information is available on declining populations among the 1,590 migrating bird species which fly from North America to Central America to winter each year, for example. It suggests that bird species are disappearing at a rate of between 1 and 4 per cent a year.[6]

4.2b Causes of extinction

There are a number of reasons why species decline and can become extinct because of deforestation.[7] An important underlying factor is related to the great diversity of species within the rainforest, which means that each species usually has few individuals and is thus very sensitive to change and stress.

Stress comes directly through the removal of habitat and associated removal of ecological niches when a patch of rainforest is burned or felled. Fragmentation of the forest ecosystem, without complete removal – as occurs in selective logging, for example, or when shifting cultivation encroaches on a rainforest remnant – also creates stress for the individuals which are removed or displaced. Habitat loss means a declining geographical range for each species, as well as increased competition for food and resources in the remaining areas suitable for habitation.

Species which remain are affected, too. A dwindling number of habitats means that the remaining species are forced to survive in a smaller area, facing greater competition from each other as well as from other species. Forest clearance might well remove all or part of the food chain for a species, which will create problems for those with specialist feeding requirements.

Deforestation affects the whole ecosystem. Each species plays a crucial role in maintaining the delicate ecological balance of the forest system, and it is an interesting case of 'one out . . . all out' when clearance starts to make forest species extinct.

Something of the nature of the biological loss is captured in the statement that

> these irreplaceable forests are the richest source of life on earth. They are home to perhaps half the world's wild creatures. Tigers, mountain gorillas, birds of paradise, rare orchids and multi-coloured butterflies are some of the unique species found only in the rainforest. The world would be a poorer place without them.[8]

4.2c Scale of losses

Before we can establish how many species are being lost as a result of rainforest clearance, we need to know how many species the rainforests contained originally. This is not without problems, because there are many different estimates to choose from.

There is a general consensus among biologists that the earth must contain something like at least 5 million species.[9] There is also agreement that tropical rainforests contain at least half of all known species on earth, and probably a great many more (see section 1.8). Rainforests are believed to contain about 70 per cent of all the plant species presently known on earth.[10] It is argued that Amazonia alone is home to an estimated 1 million species of plant and animal, including 1,600 species of birds (more than anywhere else on earth), hundreds of different tree species and as many as 40,000 species of insects.

But only about 1.5 million species are properly recorded by science today so there are a great many unknowns, particularly in rainforests. It is inevitable that most species will disappear unseen and unrecorded. United Nations Environment Programme figures for 1990 suggest that tropical species have already been reduced by 41 per cent, with most of the losses accounted for by insects.

Estimation of present rates of extinction is nearly as hard as estimating original species numbers, and the estimates also vary a great deal from one study to another. Some think it probable that one species becomes extinct every half-hour as a result of the destruction of tropical rainforest,[11] making an annual total of 8,760 species. Other estimates vary between 1 and 50 species a day world-wide (between 365 and 18,250 species extinctions per year).[12]

4.2d Future prospects

Forecasts of likely future losses are even more alarming, particularly if based on pessimistic estimates of present rates. One study suggests that if present deforestation trends continue a total of 750,000 species are likely to become extinct by the year 2000 (at a rate of over 60,000 species a year).[13] Within the next century it forecasts the extinction of 1,600,000 species, a third of all life on earth.

The ecological consequences of such massive losses of species are difficult to imagine. Yet this is no scare-mongering. Internationally respected scientists are lending their weight to the debate. For example, Peter Raven (Director of the Missouri Botanical Garden) suggests that at present rates of forest clearance

> it appears likely that no fewer than 1.2 million, at least a quarter of the biological diversity existing in the mid-1980s, will vanish during this quarter century or soon thereafter, and that a much higher proportion of the total will follow by the second half of the next century, as the remaining forest refuges are decimated.[14]

One factor underlying such pessimistic forecasts is the likelihood of what biologists refer to as an impending 'extinction spasm' or 'cascade of extinction'.[15] The complex forest ecosystem has a fragile and finely poised equilibrium, and most of its species have interlocking life-cycles and ecological requirements. Consequently the loss of one or more species can trigger the downfall of many other interrelated species, for example by altering food supplies.

Destruction of the rainforest, it is argued, 'will be the greatest biological disaster ever perpetrated by man, creating a spasm of extinction unequalled since the disappearance of the dinosaurs, over 60 million years ago'.[16]

4.2e Extinction is for ever

Rainforests which have evolved over millions of years cannot be replaced once they have all been destroyed. The loss will be permanent.

There have been times in the geological past when the rainforest has been in decline, but regeneration has always been possible because the abundance of species which survived allowed the forest to recover naturally. But recent forest clearance is very different. For a start, it is caused by human activity (thus at least in theory it is avoidable). It is much quicker than any previous natural declines, and this time it involves a wholesale loss of species.

Species extinctions mean a shrinkage of the natural gene pool from which new species might emerge, so the forest's recovery potential shrinks accordingly. Norman Myers describes the loss of genetic diversity as 'the greatest single setback to life's abundance and diversity since the first flickering of life four billion years ago'[17] – although some ecologists regard this as an overstatement (especially when the scale of mass extinctions in the Permian are taken into account).

4.3 LOSS OF NATURAL RESOURCES

Rainforests also provide a rich storehouse of wealth for people in the form of fruits, foodstuffs, timber, industrial raw materials and medicines. A wide spectrum of goods and services are available from the rainforests;[18] some examples are listed in Table 4.1.

Species extinction means the loss of these products, many of which we depend on daily (generally without ever realising it). But the loss of known resources is only part of the concern, because known and named species are just the tip of the forest species iceberg. Unknown numbers of rainforest species have not yet been discovered by modern scientists, although many are familiar to native forest peoples (see section 5.2).

Of those forest species which are known, relatively few have been examined for possible uses to mankind – less than 1 per cent according to some estimates.[19] Even then the picture is unclear because we have a tendency just to value those species which have some measurable economic potential. The International Union for the Conservation of

Table 4.1 Some goods and services available from tropical rainforests

(1) Indigenous consumption	
*fuelwood and charcoal	(cooking, heating and household uses)
*agricultural uses	(shifting cultivation, forest grazing, nitrogen fixation, mulches, fruits and nuts)
*building poles	(housing, buildings, construction, fencing, furniture)
*pit sawing and sawmilling	(joinery, furniture, construction, farm buildings)
*weaving materials	(ropes and string, baskets, furniture, furnishings)
*sericulure, apiculture and ashes	(silk, honey, wax)
*special woods and ashes	(carving, incense, chemicals, glass-making)
(2) Industrial uses	
*gums, resins and oils	(naval stores, tannin, turpentine, distillates, resin, essential oils)
*charcoal	(reduction agent for steel-making, chemicals, polyvinyl chloride (PVC), dry cells)
*poles	(transmission poles, pitprops)
*sawlogs	(lumber, joinery, furniture, packing, shipbuilding, mining, construction, sleepers)
*veneer logs	(plywood, veneer furniture, containers, construction)
*pulpwood	(newsprint, paperboard, printing and writing paper, containers, packaging, dissolving pulp, distillates, textiles and clothing)
*residues	(particle board, fibreboard, wastepaper)

Source: summarised from Myers (1988d).

Nature (IUCN) estimate that one in every six rainforest species has a non-economic use.[20]

It is useful to examine some of the major uses of rainforest species, because this reveals a lot about what is lost when forest species become extinct. The three main areas of use are as industrial raw materials, in food and farming and in medicine.

4.3a Industrial raw materials

Materials from the rainforests are used in a seemingly endless variety of ways. It is not just the tropical hardwood that has industrial value; almost everything in the forest has a use.

There is not space here to compile a complete list of important industrial products obtained from plants found in the tropical forests, but we can look at some examples. A fairly comprehensive list was compiled by the World Bank and other agencies in 1985.[21] It included oils, gum, latexes, resins, tannins, steroids, waxes, edible oils, rattans, bamboo, flavourings, spices, pesticides and dyestuffs. The list also included consumer goods made from forest products, such as wickerwork chairs, coffee, lubricants, glue for postage stamps, golf balls, chewing gum, nail varnish, deodorant, sound-proofing, toothpaste, shampoo, mascara and lipstick.

The market in these industrial products is worth billions of dollars a year. Exports of Indonesian rattan – a fibrous, spiky climbing plant used for making ropes, mats, nets and other products – is worth US$90 million a year.[22] Interestingly, such products can be exploited without destroying rainforests.

The trees provide timber and timber products which can be exported, earning valuable foreign exchange necessary for development. Many different consumer products are made from tropical hardwood, from wooden toilet seats and window frames to park benches and railway sleepers. Musical instruments such as clarinets and pianos, oboes and bagpipes are also made from tropical wood.

The most famous commercial product of the rainforest, after trees, is without doubt rubber. Even golf balls depend on the survival of the forests, because their tough, elastic outside is made from the milky latex (sap) of a tall tree (*Mimusops globosa*) found in the South American rainforests.

Forest materials are not just used by large industries, they support craft industries which provide a livelihood for many people in developing countries. For example, leaves from forest plants and trees are used to weave mats and baskets, and cane is made into furniture. Much of the craft produce is exported or sold to tourists.

Other forest products have great potential as renewable and environmentally friendly energy sources. It is estimated that 12 km^2 of fast-growing 'ipilipil' tree could provide the fuel equivalent of a million barrels of oil per year. Six Philippine 'petroleum nut' trees can produce 300 litres of oil per year for cooking or lighting.[23]

4.3b Food and agriculture

Coffee and bananas are well-known foodstuffs which originated in the rainforest, but there are many more. Forests yield many different types of fruits, cereals and nuts and half the world's main crops were originally discovered in the tropical forests.[24] Amongst the more widely used are tea, coffee, sugar, bananas, oranges and lemons, pineapples, avocados, aubergines, rice, maize, cocoa, cashews and peanuts.[25] About twelve crops provide 90 per cent of the world's food and half of them are descended from tropical forest plants, including rice and maize.[26]

It is surprising how many of the foodstuffs which we take for granted originated in the rainforest. The domestic chicken, for example, was bred from the red jungle fowl of Indian forests.[27] Many spices such as like cloves, vanilla and cinnamon which are used to make cakes and biscuits also come originally from the rainforest. More than 1,650 known tropical forest plants have potential as vegetable crops.[28]

Many of the rainforest plants are highly nutritional. The kiwi fruit (Chinese gooseberry), discovered in rainforests of South-East Asia, is 15 to 18 times richer in vitamin C than oranges. Paraguayan Indians use a forest plant which contains a chemical which is calorie-free and tastes 300 times sweeter than sugar.[29]

Much of the concern over rainforest species extinction relates to the loss of genetic variety.[30] The rainforest gene-pool has immense variety and untold potential; it is an irreplaceable natural asset to which it is impossible to attach a realistic economic value. Loss of genetic variety is much more than just a financial problem. The dozen crops which provide nearly nine-tenths of our food are susceptible to pests, disease and environmental (particularly climatic) change. It is essential to maintain genetic diversity as a safeguard for existing crops.

Preservation of genetic diversity is also essential for cultivating new plants and breeding new animals. Some rainforest species were very important ingredients in the hybridisation of new crops for the Green Revolution. Wild relatives of many commercial crops continue to provide new genetic materials to improve yields and increase resistance to pests and disease.[31]

Maize provides an interesting illustration of the need to maintain a capacity to create new crop varieties.[32] The United States is the world's largest grower of maize, which it exports to millions of people around the world. In 1970 the maize crop was blighted and half was lost. A major long-term food disaster was avoided through the discovery of an immune form of maize in the Mexican rainforest. A search began for other types of native maize, and in 1987 an important new species was discovered in a small patch of Mexican forest. This new species is resistant to at least seven major diseases and can be grown in a cool damp environment (in areas previously thought unsuitable for maize growing). World maize production could increase by up to a tenth if this newly discovered species is widely used. Such a valuable species was almost lost before it was found, however, because only a few thousand stalks remained when it was discovered in a small surviving patch of forest undergoing rapid clearance.

The tale of the maize neatly illustrates the main worry many scientists have about the prospect of losing unknown species through destruction of the world's surviving rainforests. To knowingly throw away the unknown would surely be a major act of human folly.

4.3c Medical uses

Rainforest species also provide a wide variety of materials used in medicine.

> The flora and fauna of the tropical forests hold an astonishing cornucopia of medicines, both for native peoples who can turn herbs and venoms into traditional remedies, and for industrial pharmacists who convert them into commercial drugs using extracts as raw materials or chemical blueprints.[33]

This natural dispensary of raw materials used in modern medicine includes antibiotics, heart drugs, hormones, tranquillisers, ulcer treatments and anti-coagulants.[34]

Up to a quarter of the prescribed drugs used in the United States are derived from tropical rainforest plants.[35] Forest plants supply at least 76 major drug compounds used in US prescriptions, only 7 of which could be commercially synthesised according to a 1973 survey.[36]

Some medicines from the rainforest serve invaluable functions. Leaves from the rosy periwinkle plant found in the drier tropical forests of Madagascar, for example, contain alkaloids which have been successful in treating Hodgkin's disease and childhood leukaemia.[37] Commercial sales of the drug world-wide are worth more than $90 million a year.

There are plenty of other examples. Nearly three-quarters of the 3,000 plants identified by the US National Cancer Institute as having anti-cancer properties come from the rainforest.[38] Steroids have been found in rainforest plants which help women overcome sterility.[39] Reserpine from the South-East Asian snakeroot plant (*Rauwolfia serpentina*) is commonly used in the treatment of hypertension, anxiety and schizophrenia.[40] Quinine derived from the cinchona tree in Peru is used to treat malaria, and diosgenin from Mexican and Guatemalan wild yams is a major component of the contraceptive pill used in birth control.[41] The complete list would cover several pages.

Rainforest plants also offer much promise of new treatments, particularly as cures for cancer and AIDS.[42] It is widely argued that if the rainforests are completely destroyed, cures for some diseases may never be found.

Preserving future potential is also important for the growing field of genetic engineering and other forms of biotechnology. The irreplaceable biological capital of the rainforests may prove highly significant in ways which are presently inconceivable.

4.4 LOSS OF ENVIRONMENTAL SERVICES

Rainforests are much more than just collections of plants and animals. They are complex ecosystems, in which the plants and animals interact together and with the rainforest environment (particularly the soils, climate and hydrology). Apart from maintaining suitable conditions for the wildlife within the forest, this web of interactions means that rainforests perform a series of important services which affect the environment both within and beyond the forests themselves.

As Friends of the Earth[43] describe it 'forests anchor larger natural cycles – the air, soil and water – upon which we all ultimately depend'. A South American tribal legend[44] puts it more bluntly – 'the tropical rainforest supports the sky; cut down the trees and disaster follows'. Forest clearance lets loose some of the forces of nature, and where forests have been cleared soil is eroded, floods and droughts become more frequent and weather patterns are disrupted (see page 93).

4.4a Soil changes

Deforestation seriously damages the rainforest soils through the agents of nutrient loss and soil erosion.[45] Forest soils are inherently poor in nutrients (see section 1.6) and fertility declines rapidly once crops are planted on cleared soils. Field studies have shown marked drops in soil acidity and concentrations of phosphorus, calcium and magnesium under pasture (Figure 4.1).

Nutrient imbalance is another consequence of deforestation. Nutrients are lost initially when the standing biomass (crop) is removed. Sulphur and nitrogen are lost after burning, but nitrogen is added in the ash. Subsequent losses occur through the leaching of nutrients from the soil, with up to 15 per cent of all nutrients lost during the first rainfall after trees are removed.

Soil erosion is an important consequence of deforestation in all environments, but particularly where rainfall is very high (as in the tropics). Various factors serve to minimise soil erosion under rainforest cover. The tight and multi-layered canopy cover of the rainforest vegetation effectively shields the underlying soil from the direct impact of rainfall. Humic material deposited on the soil by the overlying vegetation acts as a sponge, effectively absorbing water and cushioning the raindrops which fall. A third factor is the root holes and holes made by burrowing animals under forest cover, which allow water to be transmitted through soils with relatively little surface erosion. Once the protective vegetation cover is cleared surface runoff is increased and erosion rates usually rise dramatically, particularly on slopes.

Torrential rains and baking summers in the tropics encourage even greater erosion. The Trans-Amazon Highway, for example, is often seriously damaged by soil erosion and undermining of embankments during the rainy season, and the damage has to be repaired by the end of the next dry season.[46]

Forest clearance causes large-scale loss of topsoil by erosion, which diminishes soil productivity. Roughly 5,000 km^2 of overworked shifting

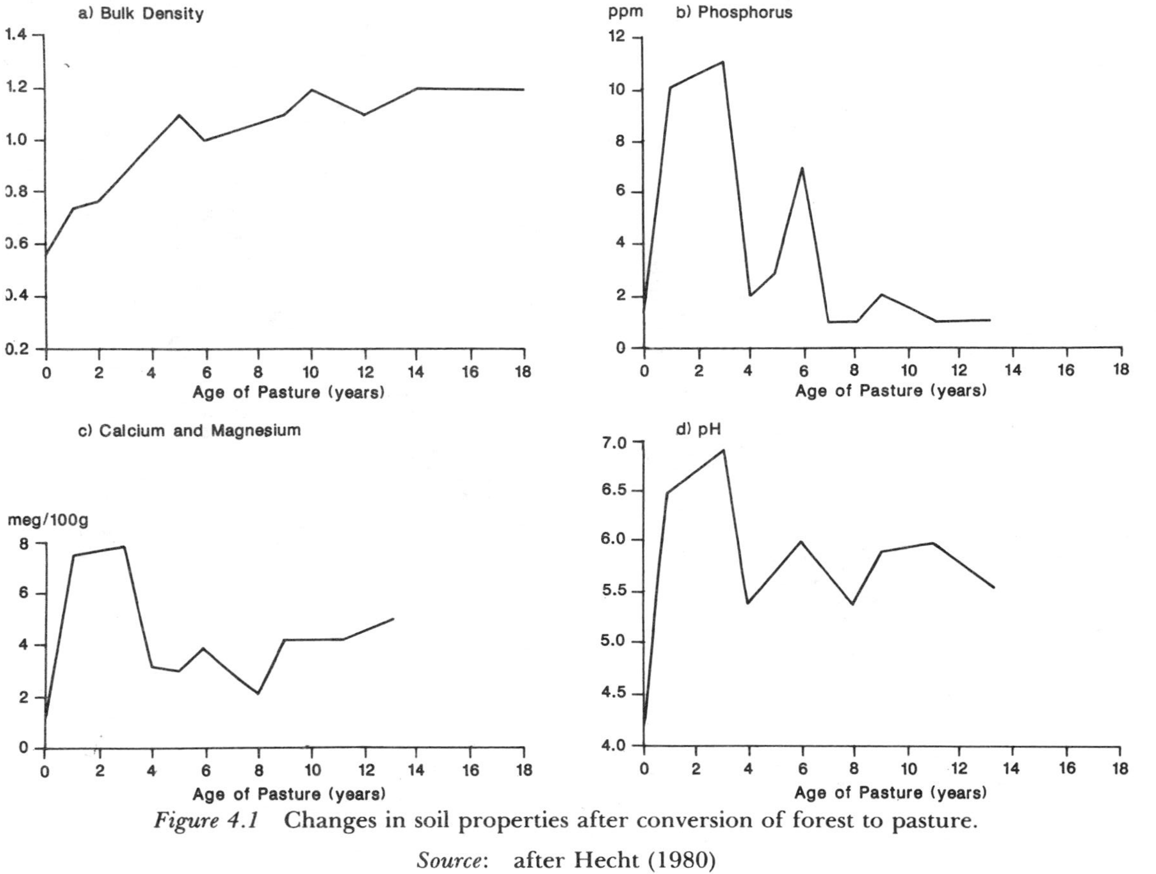

Figure 4.1 Changes in soil properties after conversion of forest to pasture.

Source: after Hecht (1980)

cultivation land in Para State, Brazil, has been virtually destroyed by widespread soil erosion.[47]

Severe erosion can strip all of the topsoil from a rainforest soil, leaving the impermeable laterite hardpan (see section 1.7) exposed on the surface. The laterite is rich in iron and aluminium compounds and it hardens when exposed to the air. Plant roots cannot penetrate through this hard shell, so the soil has no potential for productive landuse (such as farming or ranching) and natural recolonisation by trees or other plants is impossible.[48]

Rapid soil erosion often creates massive gullies, which channel surface runoff and concentrate subsequent erosion. As well as greatly increasing sediment yields in rivers downstream, such gullies can make it impossible to travel across or build on badly eroded forest soils.

The worst soil erosion is usually associated with the building of logging roads in rainforests. These are completely bare and create artificial channels along which runoff can flow. Logging activities themselves seriously damage soils, normally through compaction by heavy machinery (see section 3.6). Logging can also trigger rapid soil erosion when heavy rainfall causes mudslides. Forty people were killed by mudslides in Thailand in November 1988 which the Thai Prime Minister blamed on illegal logging.[49]

4.4b Downstream silting

The extensive soil erosion also leads to the downslope transfer of sediment to river channels, which can in turn cause silting and flooding of rivers downstream.[50] Water supplies can be contaminated, particularly where sudden massive erosion rapidly increases the sediment load to produce 'brown-outs', which is common in Ecuador, Kenya and Thailand. Over a billion people world-wide depend on water from tropical forests for drinking and crop irrigation,[51] so any reduction in water quality is likely to affect many people directly.

Rapid increases in soil erosion caused by forest clearance can leave downstream reservoirs and irrigation systems silted up, river beds clogged and farmland smothered.[52] The power-generating capacity of major hydroelectric schemes can be reduced if turbines become silted up. Some previously navigable rivers, such as the Betsiboka River in Madagascar, can no longer be used by large craft because of such deposition.[53]

A graphic example of silting caused by deforestation centres on the Panama Canal which links the Atlantic and Pacific Oceans. In the order of 9 billion litres of water are needed every day to allow thirty large passenger ships to pass through its lock system, and this is supplied by a series of reservoirs which feed the canal. One of the major reservoirs,

Lake Madden, is rapidly silting because shifting cultivators have cleared three-quarters of the rainforest from its catchment.[54] By 1988 the reservoir had lost about 5 per cent of its storage capacity, and this is expected to rise to 20 per cent within fifteen years. A loss of water supply on this scale is big enough to make the canal impassable to large ships.

4.4c Downstream flooding

Low-lying land downstream from deforested areas is often subjected to increased depth and frequency of river flooding as a result of a number of factors conspiring together.[55]

Forest clearance removes the protective vegetation cover, so more rainfall reaches the ground. Evapotranspiration losses (often in the order of 40 per cent under natural forest) fall dramatically when the rich vegetation is removed, so a greater proportion of the rainfall is available to produce runoff. The very act of forest clearance, particularly where bulldozers are used, compacts the soil. This decreases infiltration rates, so more water becomes surface runoff. Soil erosion can strip topsoil and expose the impermeable laterite hardpan, so all the runoff is surface flow (rather than sub-surface or groundwater flow). Channel silting reduces the rivers' capacity to carry large amounts of floodwaters, so over-bank flooding is largely inevitable.

The net result is that a much higher proportion of the rainfall finds its way more or less directly into river systems, greatly increasing the likelihood of damaging flash-floods. Examples can be cited from many different areas. Evidence from the Philippines suggests that widespread flooding following typhoons and monsoons in the mid-1980s was a direct result of deforestation.[56] The land area liable to river flooding in India has doubled to around 800,000 km^2 in recent years as a fifth of the country's forests have been cleared.[57] Massive flooding in China's Sichuan province in 1981 has been linked with deforestation in the upper reaches of the Yangtze River.[58]

Deforestation has caused increased flooding in the Upper Amazon,[59] where most human settlements are located along rivers and people's lifestyles are intimately related to the annual flood cycles.

4.4d Drought–flood cycle

Whilst deforestation can promote worse river flooding, the converse is also true. During periods of low rainfall there is greater potential for drought in deforested areas. A drought–flood cycle can develop in association with soil erosion (Figure 4.2) which leads to declining fertility. The cycle is initiated because forests no longer soak up rainfall

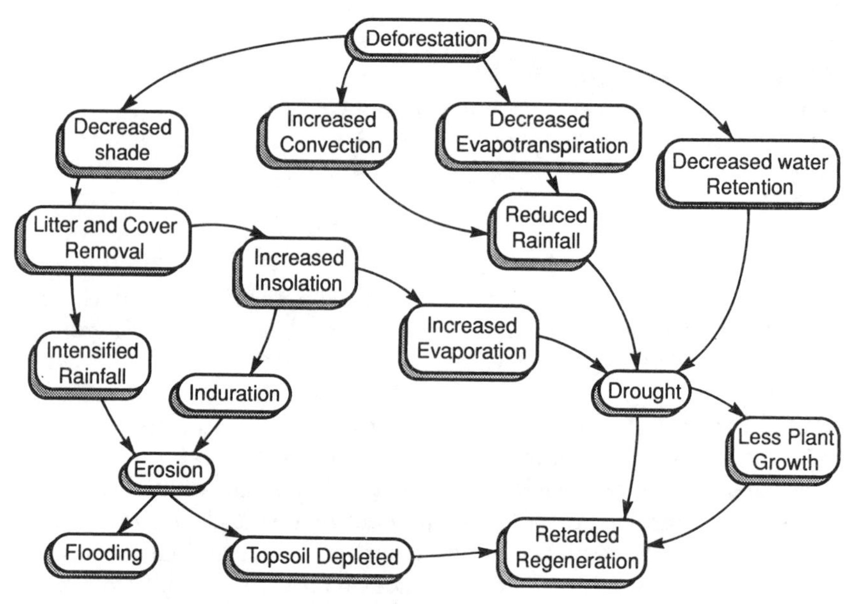

Figure 4.2 Factors in the drought–flood cycle triggered by forest clearance.

Source: after Goodland and Brookman (1977)

and use the soil as a store,[60] and water shortages can be serious. But heavy rains can quickly bring floods to replace the droughts.

Drought–flood cycles associated with forest clearance appear to be particularly common in the valley lands of Southern Asia. Farmers there can no longer rely on rivers such as the Ganges, Brahmaputra, Irrawady, Salween and Mekong for regular water supplies.

4.5 LOCAL AND REGIONAL CHANGES TO CLIMATE

Clearance of the rainforests can affect climate in a number of ways and on different scales.[61] Local and regional climates are affected mainly by changes in surface albedo (that is, reflectivity of the ground surface) and changes in local hydrology (particularly in evapotranspiration and rainfall). These tend to raise temperatures and lower rainfall, resulting in more regular and more serious droughts.

4.5a Albedo and temperature

Short-wave incoming radiation from the sun passes through the atmosphere and reaches the earth's surface, which it heats by absorption. When it bounces back off the earth's surface it is converted into long-wave re-radiation energy, which then heats the atmosphere. The conversion process, thus the heating of earth and sky, depends mainly on the nature of the surface, such as whether it is land or sea, and what type and character of vegetation covers the land.

Reflectivity is normally measured as the 'albedo', which expresses the percentage of radiation which is reflected back into the atmosphere rather than absorbed by the earth. The lower the albedo, the greater the amount of radiation absorbed and thus the smaller the heating of overlying air.

Light-covered and bright surfaces such as desert sands have high albedos (around 40 per cent), whereas dark and dull surfaces such as forests have low albedos. Tropical rainforest, with its dark evergreen canopy, has among the lowest albedos of any vegetation cover, as low as 9 per cent in Kenya.[62] But clearance of the forest raises the albedo because soils and crops have higher reflectivity.

A rise in the albedo means greater heating of the local atmosphere, thus local warming is likely. If many large patches of rainforest are cleared, it is feared, local warming could extend to a whole region.

Regional warming on this sort of scale is by itself unlikely seriously to affect global circulation patterns and thus temperatures. But there are prospects of influencing world climate through deforestation because the clouds above the rainforests absorb large quantities of solar radiation. Forest clearance decreases cloud cover (because evapotranspiration rates fall; see section 4.5b), which could lead to an increase in the earth's albedo. Such a change could trigger a chain-reaction involving disruptions to atmospheric convection patterns, wind currents and rainfall on a global scale.

Some of these possible consequences of deforestation are speculations based on principles of climatology. As yet there is little detailed scientific evidence to support or reject them. One of the few empirical studies which have been conducted concluded that the temperature changes triggered by albedo changes are likely to be minimal.[63]

What is more certain is that soils which were previously kept relatively cool by the forest shade are heated after deforestation. Studies in Guatemala, for example, have shown that as little as 4 per cent of solar radiation reaches the soil surface under forest cover, and twenty-five times more radiation reaches bare soil.[64] Heating of the forest soils speeds up the natural processes of hardpan formation (see section 1.7) and nutrient leaching (see section 1.6) which quickly make them useless for productive use (Figure 4.3).

Deforestation is also known to bring more extreme local variations in air and soil temperatures. Under forest cover diurnal variations in surface temperature are moderated because air is effectively trapped beneath the forest canopy, shaded from sunlight. Clearance can disrupt

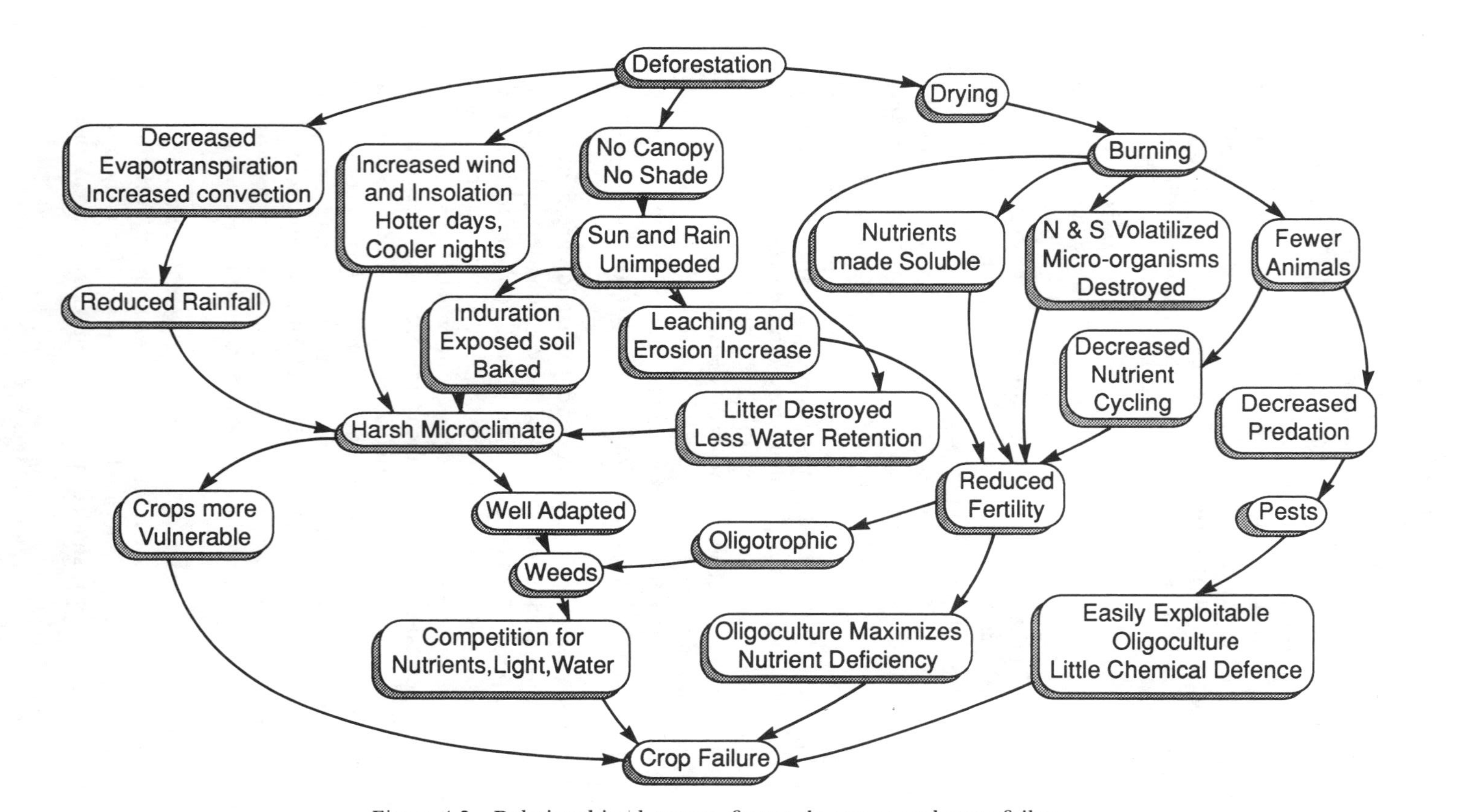

Figure 4.3 Relationships between forest clearance and crop failure.

Source: after Goodland and Brookman (1977)

local weather patterns by warming air which was previously kept cool and producing surface conditions which are hotter during the day and colder at night.[65]

Deforestation can also cause temperature changes indirectly, via localised changes in atmospheric moisture. Trees perform a very important function in transferring heat from the earth to the atmosphere through evapotranspiration – heat is released as water vapour from the trees, cools and condenses. Removal of the trees also means removal of this heat-transfer function.

4.5b Evapotranspiration and rainfall

Closely associated with the impacts of albedo change are the consequences of deforestation on evapotranspiration and the regional hydrological cycle.

Evapotranspiration, a combination of evaporation and transpiration, is the main mechanism by which moisture and energy are transferred between land and the overlying atmosphere. Trees are particularly effective in recycling a large proportion of the rainwater which falls over them, drawing it up from the soil below via their roots and then releasing it as water vapour through their foliage.

Evapotranspiration rates are higher in rainforests than in any other type of vegetation, because of the high rainfall, high temperatures and closed canopy (which creates very humid hot-house conditions). This moisture condenses to form thick cloud cover in the atmosphere above the forests, and much of it falls directly back over the forest as convective rainfall (particularly during intense storms; see section 1.2).

This unique combination of high rainfall, high evapotranspiration rates, limited air circulation and thick cloud cover means that the normal water cycle is effectively short-circuited over rainforests. Most of the rainforest downpours usually come directly from forest-generated rain.

The Amazon Basin, which contains two-thirds of all the freshwater on earth, illustrates this short-circuit process clearly. Studies have shown that the Amazon 'with easterly winds and the barrier of the Andes . . . has an almost unique water recycling regime'[66] in which more than half (possibly up to 80 per cent) of the rain which falls on the forest is recycled to fall on the same area repeatedly. A typical turn-over period is five days.[67]

Evidence of reductions in local rainfall which have been attributed to clearance of tropical forests is available from many areas, including India, Malaysia, Philippines, the Ivory Coast and Costa Rica.[68] One consequence of such reductions in local rainfall is that dry periods will be extended in these areas. Ironically such environmental stress might be serious enough to accelerate the natural death of surviving rainforests there.[69]

It is generally agreed that deforestation leads to reductions in convective rainfall, but there is less agreement on why this occurs. At least three sets of factors and mechanisms are involved. The first is the reduction in evapotranspiration when trees are replaced by bare soil or crops. This means less cloud cover and thus less atmospheric moisture overhead, so the short circuit is removed.

A second factor, closely tied to the first, is that forest clearance decreases infiltration and encourages more surface flows of water and greater streamflow in rivers. This means that any available water is rapidly transported away from the deforested area, leaving less available for evapotranspiration and the formation of clouds overhead.

The third proposition is based on the effects of land surface roughness on the distribution of convective rainfall.[70] It is well established that vegetation roughness causes a drag on winds crossing an area, which dissipates some of the winds' energy. Loss of forest cover means a more streamlined ground surface with less roughness. Wind energy (now no longer constrained by the rough forest cover) is thus maintained, which may promote decreases in rainfall. It may also bring more frequent and more intense storms capable of destroying natural vegetation and agricultural crops and causing serious soil erosion.[71]

Regardless of the precise mechanisms involved, there are fears that deforestation might disturb rainfall over areas much wider than just the centres of clearance. It may ultimately create serious regional problems. Meteorologists have speculated whether rainfall cycles over much of the northern hemisphere might even be disrupted by tropical deforestation, and concern is growing that Amazonian clearance might disturb the movement of moisture-bearing air-masses which regulate weather patterns far away in North America's prime agricultural areas.[72]

4.5c Aridity and drought

Deforestation can set in motion a chain of events which significantly increase the real threat of drought.[73] Increased runoff, reduced evapotranspiration, declining soil water storage, increased temperatures and reduced precipitation can quickly lead to a cycle of drying and the progressive onset of drought.

The reduction in cloud cover (see section 4.5b) is a significant link in the chain, because it decreases the prospect of rainfall and increases the proportion of incoming solar radiation which reaches the ground surface and thus guarantees increased heating. This particular link can affect areas great distances downwind of the deforestation, which receive less rainfall and can suffer from increasing aridity and even face drought.

The problem is most acute in tropical areas which already experience a severe dry period naturally. Any decrease in rainfall will prolong the dry period, put vegetation and crops under severe stress and cause serious water shortages. Parts of southern Brazil are believed to be facing a critical increase in aridity because of deforestation, which threatens to make natural recovery of the rainforest environmentally impossible (even if clearance were stopped and massive replanting began) (Figure 4.3).

It is even argued that the serious prolonged droughts which have inflicted so much human misery and suffering in the Sahelian Belt of North Africa are partly caused by the destruction of West African rainforests.[74] Forest clearance is believed to set in motion a spiral of problems. When trees are felled local supplies of fuelwood become scarce, so people burn crop residues and animal dung for cooking and heating. This deprives the already impoverished soil of nutrients and declining soil fertility causes crop failure. The grassland (previously forest) is over-grazed by domestic herds, which reduces it to desert. Reductions in cloud cover, increased temperatures and reduced rainfall serve to intensify an already arid and precariously poised situation. Impoverished people move on to other wooded areas and repeat the process.

4.6 GLOBAL CLIMATIC CHANGE

Deforestation can also trigger global climatic change by altering atmospheric circulation patterns (via changes in the earth's latent heat flux) and altering atmospheric chemistry (via changes in greenhouse gases). These tend to have different impacts in different places (particularly at different latitudes), but the suspected net effect is global warming.

4.6a Heat flux

The reductions in evapotranspiration and cloud cover in the tropics, caused by deforestation, could have world-wide consequences because of what is known as the global latent heat flux.[75] This is the transfer of heat and energy from low (hot) latitudes to high (cooler) latitudes which maintains thermal equilibrium on earth by using tropical heat to compensate for polar cold. In technical language, the positive radiation flux of the tropics balances the negative radiation flux of the poles.

Under normal conditions heavy cloud cover is formed in the tropics by evapotranspiration, much of it from the rainforests. These clouds reflect incoming short-wave solar radiation, so they prevent overheating in the tropics. Convection currents in the troposphere (lower atmosphere) carry the tropical clouds to higher latitudes where the water vapour falls as rain (which is warm relative to the cooler atmosphere that far north or south from the equator).

In this way the evapotranspiration within tropical rainforests plays a critical role in maintaining thermal balance in the earth's atmosphere and preventing a significant increase in temperature differences between different latitudes.[76] Any marked reduction in evapotranspiration and cloud cover in the tropics could thereby trigger widespread changes in climate.

The likely climatic impacts can be predicted from first principles.[77] Tropical deforestation gives rise to reduced evaporation and reduced convective activity and rainfall. This causes a reduction in the release of latent heat from the tropics, which weakens the Hadley circulation. The net effect is to increase precipitation in the low latitudes (between 5 and 25°N, 3 and 25°S) and reduce the precipitation gradient between the equator and the poles. A weakening of the Hadley cell also reduces the transport of heat and moisture from the equator towards the poles. In higher latitudes (between 45 and 85°N, 40 and 60°S) this means lower temperatures and less precipitation.

The 'altered heat flux' scenarios anticipate warmer, wetter tropics and cooler, drier temperate areas arising from tropical deforestation. But they ignore the global warming which might be caused by changes in carbon dioxide levels in the atmosphere, via the 'greenhouse effect' (see section 4.6c).

4.6b Pollution filter

Global climate might also be affected by deforestation through the loss of a valuable natural pollution filter which trees provide.[78] Trees produce oxygen and take in carbon dioxide (CO_2) by photosynthesis.

Deforestation, by removing this natural air conditioning system, might have two consequences. Trees purify the air we breathe and forests play a significant role in maintaining the oxygen balance of the earth. Clearance might mean a shortage of oxygen for life on earth to survive.

The second and more damaging effect stems from the very effective role which forests play in filtering carbon dioxide from the atmosphere. Rainforests act as a carbon sink and prevent the build-up of CO_2 in the atmosphere, acting as the 'lungs' of the earth. This helps to constrain global warming triggered by greenhouse gases.

Fears have been expressed that forest clearance is eroding this natural pollution filter and thus removing the check on global warming.[79] Clearance by burning, which is very widespread, amplifies the problem because large-scale wood burning will deplete oxygen in the atmosphere and release more carbon in the form of CO_2 (which will promote global warming).

The significance of this loss of pollution sink is widely debated, and there is little hard evidence to confirm or reject it. Some take the threat seriously whereas others dismiss the prospect as 'just a myth'[80] which is much less important than other mechanisms of climatic change.

4.6c Greenhouse gases

The clearing and burning of tropical forests are believed to have made a significant contribution to recent warming of the earth's atmosphere via the so-called 'greenhouse effect'.[81] The heating comes about through greenhouse gases which trap long-wave radiation reflected back into space from the earth's surface.

The gases operate in much the same way as the glass of a garden greenhouse. Energy coming from the sun as short-wave radiation passes through the ozone layer in the atmosphere where harmful elements are filtered out. This incoming radiation hits land or sea and is either absorbed as heat or reflected back into space as long-wave radiation. The greenhouse gases are virtually transparent to incoming (short-wave) solar radiation but absorb outgoing (long-wave) terrestrial radiation. They trap the long-wave radiation, thus warming the lower atmosphere.

The four main greenhouse gases are carbon dioxide (CO_2), methane (CH_4), chlorofluorocarbons (CFCs) and nitrous oxides (NO_x), although only the first two are believed to be significantly increased by tropical deforestation. Carbon dioxide accounts for roughly half of all greenhouse gas emissions, so it is a major contributor to global warming. It is also the most important greenhouse gas associated with tropical deforestation.

Most of the greenhouse gases are produced as air pollution, principally in the industrial areas of the developed countries. Destruction and burning of tropical forests[82] are the second largest cause of increased atmospheric CO_2.

Rainforests store more carbon in their plant tissues, decaying material and soil than other vegetation types. When forest is cut and burned (the fate of most cleared patches), the carbon is returned to the atmosphere as carbon dioxide in the smoke from the fire. Clearance also reduces the amount of CO_2 that can be removed from the atmosphere by photosynthesis (see section 4.6b). The net effect is a progressive build-up of CO_2 in the atmosphere, to add to that produced by air pollution.

Direct measurements are impossible, but calculations suggest that deforestation injects between 5 and 10 billion tonnes of CO_2 into the atmosphere each year, between 3 and 6 billion tonnes of it coming from the tropical forests.[83] Something like a fifth of the CO_2 recently added to the atmosphere is believed to be a result of deforestation.[84]

One study suggested that 512 million tonnes of CO_2 were injected into the atmosphere as a result of forest burning in Amazonia during 1987.[85] More recent evidence suggests that deforestation during the late 1980s was releasing in the order of 1.4 billion tonnes of CO_2 a year.[86]

This is an addition to the atmospheric carbon dioxide budget and thus an important contribution to the greenhouse effect. It has been estimated that if all the world's rainforests were burned between 1986 and 2000, the CO_2 content of the atmosphere could rise by between 15 and 20 per cent.[87] Norman Myers[88] estimates that clearance of tropical forests is contributing about 30 per cent of the build-up of CO_2 in the atmosphere, which in turn accounts for roughly half of the global warming.

But the impacts of deforestation are not confined to CO_2. Nitrous oxide (NO_x), which is believed to be much more effective than CO_2 in trapping heat within the atmosphere, is also released following forest clearance but there is little hard evidence of rates or quantities.

Methane (CH_4), which accounts for about a fifth of the total emissions of greenhouse gases, is one of the fastest-growing gases. It is about twenty times more effective than CO_2 at absorbing outgoing infra-red radiation in the troposphere.[89] The main sources are rice paddies, belching cattle and biomass burning (especially areas of cleared rainforest). Cattle ranching in tropical forest areas contributes to the methane emissions. It is even argued that dam construction and subsequent flooding might contribute to the methane emissions, with methane produced by alternate flooding and exposure of large land areas.[90]

4.6d Global warming

Debate continues about what the consequences of greenhouse warming are likely to be, both globally and in particular areas. The debate gets even more contentious when it comes to trying to apportion damage to the various contributory factors (for example, how much global warming might be caused by air pollution and how much by tropical deforestation).

Many scientists predict that atmospheric CO_2 levels will double by the year 2050.[91] Such an increase would mean that an extra 1.5 per cent of solar radiation would be trapped in the earth's atmosphere, causing an increase in average air temperatures of between 2°C and 3°C.

Such warming might increase evapotranspiration in many areas outside the tropics, which would in turn increase cloud cover, thus inhibiting further temperature rises. But in the tropics evapotranspiration is likely to fall, producing less cloud cover and a significant increase in desertification.

Equatorial regions are likely to end up much worse off if global warming, triggered by the greenhouse effect, occurs to a significant degree. In the long term the tropics might not be suitable for sustained growth of rainforest, if any were to survive the clearance pressures.

Other projected consequences of global warming are far-reaching and well documented elsewhere.[92] There would be relatively few winners in this game of environmental Russian Roulette. Some high latitude areas which were previously too cold for extensive arable farming may be able to increase crop yields, and places like India and the Middle East could become wetter and more fertile.

But many more areas would be losers. The earth's main fertile regions are likely to become drier and less productive. Many areas should expect much larger seasonal temperature variations and more erratic weather patterns. Tropical storms are likely to become more frequent and more violent. Thermal expansion of the warmer seawater, coupled with melting of polar ice caps, will probably cause a rise in sea level putting many low-lying areas under serious risk of permanent flooding. Some areas would be suffering from drought while other areas (even nearby) might be under floodwater.

4.7 CONCLUSION

There is clearly a great deal more at risk from tropical deforestation than just trees. The stakes are high, and the consequences are likely to affect people throughout the world. Species extinction means throwing away invaluable and often unknown genetic reserves, destabilising other ecosystems, shrinking the world's biological capital assets and abandoning any possible future use of rainforest resources in farming, medicine or industry.

Destroying the environmental stability of rainforest areas and promoting soil erosion, river flooding, silting and other hydrological changes will ensure that vast areas downstream from the deforestation are put at risk too. Forest clearance threatens climatic change on a variety of scales, including irreversible changes to temperatures and rainfall.

Underlying most of these anticipated impacts of deforestation is considerable uncertainty and speculation. As one recent study of the regulatory role of rainforests on global weather patterns concluded, 'their role in maintaining the global climate is poorly understood, but removing the forests must be the worst possible way of finding out exactly how crucial they are'.[93]

5

FOREST PEOPLES

The voice of the people hath some divineness in it,
else how should so many men agree to be of one mind?
Francis Bacon, *De Dignitate et Augmentis Scientiarum* (1640)

5.1 INTRODUCTION

Although deforestation is having serious impacts on plant and animal species, local environmental systems and regional if not global climate (Chapter 4), by far the most direct, immediate and long-lasting impacts are on native forest peoples.

5.1a Human casualties

The millions of indigenous people in over seventy tropical countries who live in and rely on the rainforest are the real casualties of forest clearance. They bear the real force of the social and ecological costs of clearance, while rewards (which are almost entirely financial) go to a relatively small minority of sponsors and investors around the world. Many are moved outside the forests to new environments to which they are neither accustomed nor suited; others are killed or die directly.

These human losses are very significant, although they rarely appear as separate items on the overall balance sheet of forest clearance. But there is also a major cultural loss because forest people take with them their traditional knowledge of forest resources and sustainable techniques of husbandry accumulated over many generations.

The number of people involved is much higher than one might think, because the problems of clearance are not confined to a handful of groups or tribes 'left behind' by modern society. Estimates from the World Resources Institute[1] suggest that the lives of more than a billion people (one-fifth of humanity) are already 'periodically disrupted by flooding, fuelwood shortages, soil and water degradation and reduced agricultural production caused directly or indirectly by the loss of tropical forest cover'. The ecological, environmental and climatic problems outlined in Chapter 4 have a disproportionately high toll among populations in the tropics.

Whilst the plight of many of these human casualties of forest clearance largely remains hidden from view in the developed countries, concerted efforts are being made to highlight their problems and mobilise public and political sympathies. Groups like Survival International and Friends of the Earth have made rainforest peoples the focus of special, high-profile campaigns. Concerned individuals like the rock star Sting[2] (founder of the Rainforest Foundation and staunch defender of the Yanomami Indians in Brazil; see section 5.5) try to keep the plight of the forest peoples under the media spotlight. A prominent eco-activist with a keen eye for the high-publicity stunt is the photo-journalist Bruno Manser,[3] who lived with the nomadic Penan tribe in the rainforest of Sarawak for seven years and compiled seven journals recording details of their language, medicinal plants and practices, food, cooking, customs, hunting and fishing methods, crafts, religion and beliefs.

5.1b Shifting fortunes

One of many paradoxes surrounding clearance of the rainforests is the way western attitudes towards the forests and their people have changed through time.

From the sixteenth century onwards, those who 'discovered' the rainforests and others who visited and explored them brought back tales of luxurious and fertile forests guarded by hostile and backward people. Many myths grew up about the so-called savages, noble or otherwise, living in the rainforests. The nineteenth-century view of these forest people as primitive savages, there to be tamed and obvious targets for zealous missionaries, survived well into the present century.

Within the past two decades or so attitudes have changed dramatically, and forest peoples are now widely seen as threatened rather than a threat. A number of factors have helped to bring this about. One is that we now have much better and more realistic ideas about indigenous peoples than we had before. More rational television documentaries and printed articles have helped to correct fanciful popular misconceptions based more on folklore and fear than on anthropology and awareness.

Another important factor is undoubtedly the dramatic change in fortune which tribal peoples now face. They are being pushed to the edge of extinction, and public sympathies are swinging in their direction.

The plight of the forest people might also have touched some raw nerves of rustic nostalgia in the industrialised west. It is easy to paint a rather sentimental (but inappropriate) image of tribal people whose lives depend on the forests, who have lived in harmony with their environment for thousands of years and whose future survival is now threatened. A cultural concern complements the humanitarian concern, because the death of a people also means the death of their culture.

The lives and livelihoods of forest peoples are also part of the set of implicit values (see section 1.9) which are often attached to the rainforests, too. Many in the west argue that these indigenous people have a right to exist and continue existing quite simply because they are there.

Public sympathies have also drifted towards the people of the forests because the pressures they face are ones imposed upon them, principally by industrialised developed countries. This raises fundamental questions of ethics, morality and rights, particularly the primary question, what rights do outsiders have to disrupt or even eliminate the lives and livelihoods of native peoples? That is part of a much broader debate over human rights and responsibilities which characterises the last quarter of the twentieth century.

5.2 TRIBAL PEOPLE AND THE RAINFOREST

Tribal peoples have lived in the rainforests for many generations and they are an integral part of the complex forest ecosystem, perfectly adapted to it.[4] They work with the rainforest rather than against it, unlike industrialised societies which tend to see their environment as something to be tamed, exploited and privatised.

Forest people have learned to live in harmony with their environment and have developed lifestyles which make use of forest resources while at the same time respecting the limited carrying capacity of the rainforest environment. There are three particular ways in which this harmony is maintained under natural conditions – through physiological adaptation to environment, through their attitudes to nature (world-views) and through their sustainable practices and use of forest resources.

5.2a Physical adaptation

Like all other forest life forms, forest people have physical adaptations to their natural home. Generations of minor physiological changes, each improving the individual's ability to survive under the demanding hothouse conditions of the rainforest, have produced some remarkable responses in their body dynamics.[5] Forest people have metabolic rates and body structures which allow them to generate less heat than other people, and to dissipate the heat they do produce more rapidly to prevent overheating. They also produce relatively little sweat, so they need less water to survive. Furthermore they only need a low intake of protein and their bodies are able to store proteins for weeks.

Another striking characteristic is that 'when they cease to be healthy, forest people die'.[6] They do not suffer from the chronic illnesses (like cancer, hypertension and heart disease) which affect so many people in the affluent developed countries. This also means that forest people have little if any immunity to the myriad contagious viruses and other illnesses so endemic in the west; ailments such as influenza can quickly kill them.

5.2b World-views

Indigenous rainforest peoples also have markedly different cosmologies or belief systems to those common in developed countries.[7] The dominant western anthropocentric (people-centred) world-view is alien to them. To them the natural world is what matters; it is spirit-filled and they are a part of it in the same way that the trees around them are. The Turkano from north-west Amazonia, for example, see nature *not* as a physical entity apart from people, so people cannot confront it or oppose it; it is part of them and they are part of it.

This world-view gives the forest peoples different sets of views, values and priorities to what is more common in industrialised societies. It also gives them a different set of ideas of the world about them. To them the forest offers a home, sustenance and protection, whereas to outsiders it harbours unknown danger and difficulty.

There are some interesting variations in perception between different forest people, such as the Pygmies and the Bantu of the African rainforest.[8] The Pygmies have flourished in the forest for thousands of years. They respect it and only feel completely free and safe within the forest. The Bantu, on the other hand, have only lived in the forest for a matter of centuries. They fear it and see it as a world of evil spirits which they must destroy and tame.

5.2c Sustainable practices

People and rainforest depend on each other because the human activities give to as well as taking from the forests. The best illustration of this symbiotic association is traditional slash-and-burn farming (see section 3.3), which allows sustainable and productive use of the poor forest soils. In return, regular clearance and revegetation of small forest patches has contributed to the huge diversity of forest habitats and species (see section 1.8).

Experience of living in and off the forest over generations has endowed the indigenous peoples with an intimate and highly practical knowledge of the forest, its abundant ecological resources and their many potential uses.

Many forest plants are used for different purposes, some collected wild and others grown deliberately. In northern Thailand, for example, the Lua tribe grow up to 75 food crops, 21 medicinal plants, 20 plants for ceremony and decoration and 7 for weaving and dyes in their complex cultivations.[9] Traditional healers in South-East Asia use 6,500 plants to cure malaria, stomach ulcers, syphilis and other disorders.[10] The Hanunoo people of the Philippines can separate 1,600 plant species while professional botanists working in the same forest can only distinguish 1,200.[11]

No forest resource is over-exploited for short-term gain.[12] This is a sensitive, well-adapted and highly sustainable lifestyle, evolved over generations and born out of practical experience blended with respect for their fellow creatures. Traditional people know far more than we will ever know about the different uses and properties of forest plants and wildlife. Such native knowledge is at risk as tribal groups are dispersed or reduced through forest clearance. As with species extinction (see section 4.2), once such knowledge is lost it is gone for ever.

Most tribes in Amazonia have devised cultural controls for limiting their populations to levels well below the carrying capacity of their rainforest environment. These include infanticide, extended periods of lactation with taboos on sexual intercourse and even the occasional murder of related tribespeople from neighbouring villages.[13]

5.3 DECLINE AND FALL

Any damage to the forest system also damages forest peoples. Most indigenous forest populations have shrunk since the arrival of outsiders and the growth of exploitation. Around 5 million native Indian people were living in the Amazon Basin in 1500, according to recent estimates.[14] By 1900 this had fallen to around one million and by the early 1980s there were fewer than 200,000 (Figure 5.1).[15]

It is impossible to know for certain how many forest people have been lost, either through history or in recent years, because tribes may disappear before they are discovered by the outside world and there are no reliable census data for known tribes. A widely quoted figure is one tribe lost in Brazil every year since 1900,[16] which may well be at least of the right order of magnitude. Certainly the evidence which does exist shows that many surviving groups and tribes are shrinking fast.

The decline of the Waimiri-Atroari Indian tribe in Brazil is better recorded than many.[17] In 1903 the population stood at around 6,000 but many died in violent encounters (often with outsiders) and by 1973 it had fallen to 3,500. There were only 374 remaining in 1986, and most of these were children. Over 3,000 Indians perished in less than twelve years. Many died of measles during epidemics; others were shot by

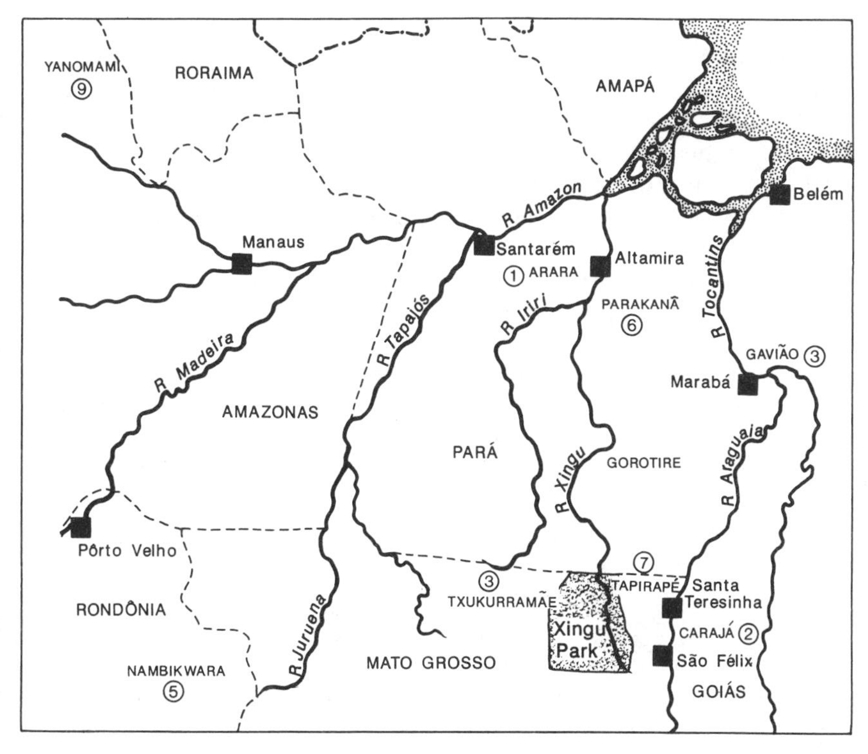

Figure 5.1 Location of rainforest Indian tribes in Brazil.

Source: after Branford and Glock (1985)

adventurers and gunslingers hired by large landowners. Displacement from tribal lands by large development schemes (see section 3.9) has brought added pressures.

Amongst the best known, thanks to world-wide publicity engineered by Sting and the Rainforest Foundation, are the Yanomami Indians from Brazil who until recently were the largest undisturbed tribe in the Americas (see section 5.5).

5.3a Culture, rights and survival

The problem is not confined to the death of tribal peoples, because survival by dispersion creates its own problems. Many forest people suffer acute culture shock when confronted with modern society, which makes it very difficult for them to assimilate easily when they are displaced from the forest and forced to resettle outside.

Death and dispersion of forest peoples caused by wholesale deforestation both bring about cultural extinction or ethnocide. The loss is potentially immense, because with the people go their knowledge and understanding of the forest, its resources and the best ways of using both. Cultural death spells the end of that branch of cultural knowledge and folk medicine.

Tribal rights are widely overlooked for surviving groups or individuals from the displaced forest tribes, too. This brings fundamental hardship to those who survive, whose cultural roots are severed and cannot be properly replaced by western culture. But it also speeds up the pace of cultural extinction, by denying tribal members the right or generally even the opportunity to practise traditional customs in traditional areas.

Tribal people's rights have been violated just about everywhere that rainforests have been cleared. In Sarawak, eastern Malaysia, for example, logging companies destroyed ancestral graves of the nomadic Penan tribe and flattened forest gardens and fruit trees.[18] The Penan were offered little or no compensation and applications to have their lands declared communal forest reserves have been turned down or ignored (see section 5.7b).

5.4 THREATS AND PRESSURES

It would be wrong to think that all tribal people living in rainforests are deliberately killed by those who are claiming and clearing the forest – the timber companies, highway constructors, ranchers and so on. They are subjected to a wide variety of different pressures, some of which kill them (directly or indirectly) while others force them to move elsewhere.

Where pressure is localised or sporadic, and large areas of fairly natural rainforest survive, forest people can remain in the forest. But they have to adapt to changing circumstances, which is not easy for them. As clearance claims ever bigger portions of the forest, the tribal people have a shrinking area in which to survive.

Many of their traditional natural sources of fuel, food, dyes, medicines and so on might disappear in the clearance, forcing them to seek out alternatives. With clearance they have to walk further for food, and hunting wild forest animals becomes very difficult. Tribes which cultivate crops in forest clearings have to farm those areas more intensively, without the familiar benefit of being able to move on and open up new areas for cultivation. A downward spiral of more intensive and less sustainable farming by native peoples can very quickly be set in motion.

5.4a Disease

Tribal groups are confronted with other pressures, too. One of the most damaging is the arrival of new populations (including peasant settlers from elsewhere, loggers, ranchers, and road and dam construction teams) who bring new diseases like influenza into the rainforest. Indigenous people normally have no natural resistance or immunity to such infectious diseases, so they are highly susceptible to them. Measles kills many tribal people, such as the Waimiri-Atroari in Brazil (see section 5.5). Up to half of the Yanomami tribe living in northern Brazil were wiped out in a matter of weeks in 1977 when an epidemic of measles was brought in from outside.[19]

Malaria is another growing problem to which forest tribes have little or no natural resistance. Between 5 and 15 per cent of the tribes in Amazonia are believed to suffer from malaria (probably imported from north-east Brazil) at any time, putting them out of action for up to a month at a time. In 1987 there were a quarter of a million cases of malaria in Rondonia alone, affecting a fifth of the total population.[20] Other infectious diseases which ravage tribal groups include Chagas' disease (a form of sleeping sickness), scrub typhus and schistosomiasis.[21]

5.4b Displacement

Displacement is a more typical fate for forest people faced with clearance by outside agencies. Sometimes this happens because tribal lands are taken away from them and used for other purposes. In Brazil, for example, the Balbina Dam (see section 3.9) has caused displacement of about a third of the Waimiri-Atroari tribe.[22]

Often displacement is forced upon an unwilling tribe by an obligatory resettlement scheme. This often involves moving them to a place which is unfamiliar to them, where they can no longer support themselves. It is ridiculous to assume that forest people can simply uproot their lifestyles and experiences and then plant them somewhere else, often in an alien environment, with no serious cultural or human damage.[23] In Indonesia the Dyak people have had to contend with wholesale disruption after a government-led campaign of forced resettlement, in which traditional dwellings were destroyed.[24]

5.5 THE YANOMAMI OF BRAZIL

5.5a Traditional communities

The Yanomami of Brazil are the last remaining large group of forest Indians in Latin America.[25] They live in the state of Roraima (which is roughly the size of Portugal) in the extreme north of Brazil, on the border with Venezuela (Figure 5.2).

This land has been their home for up to 10,000 years. Like all tribal peoples they have lived off the forest without destroying it, and they

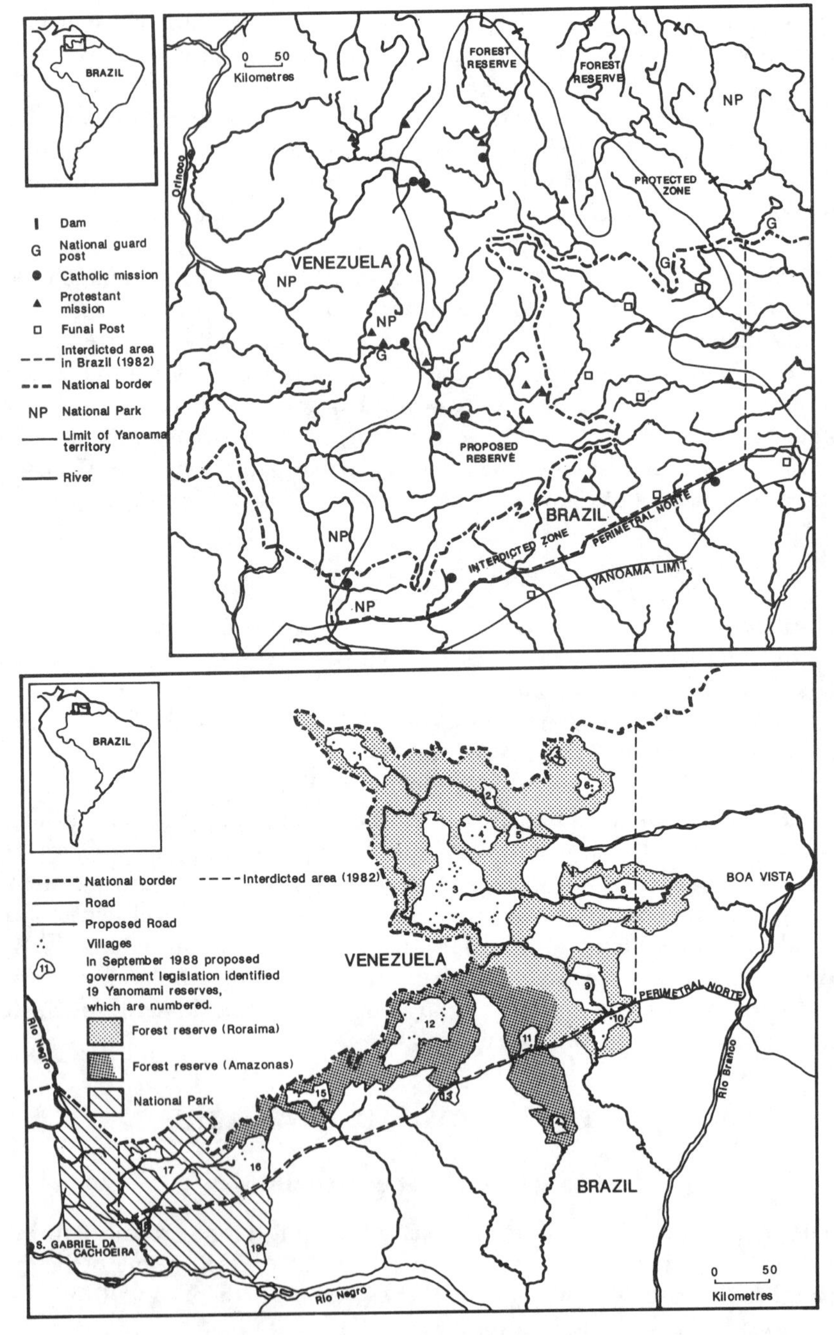

Figure 5.2 Yanomami lands and forest reserves in north-west Brazil.

Source: after Johnson *et al.* (1989b)

have an unrivalled understanding of the medicinal and nutritional resources of the rainforest. They are typical forest folk who hunt large animals, fish and birds and cultivate a wide variety of wild plants in small gardens. Something like 2,000 wild plants are collected from the forest over the year and used for different purposes.

A few Yanomami communities, isolated by mountains in the west, still had little contact with the *nabe* (outsiders) even at the close of the 1980s. Others, in less remote areas more accessible to the *nabe,* were proud owners of shotguns, fish-hooks and torches (used mainly to hunt alligators at night).

But in most other ways the Yanomami have retained their tribal habits, customs and practices. They live scattered throughout the forest in *malocas,* circular thatched huts each housing between 10 and 20 families.[26] Each communal shelter is an autonomous village, related to its neighbours by marriage, trade and festivals. Communities move every 5 to 10 years when game becomes scarce and poor soils in the gardens are exhausted.

Until 1986 the Yanomami were more or less undisturbed by outsiders, but since then there have been sweeping changes. The two most serious pressures confronting them have been invasion by gold prospectors and subdivision and fragmentation of their traditional tribal lands by government decree.

5.5b Gold rush

Roraima had been a peaceful and relatively remote area of rainforest until 1987. Government studies published that year revealed the presence of gold, diamonds, tin and bauxite in the state's Indian lands, and towards the end of the year a smallholder struck gold in the north-west corner of the state. His instant wealth triggered a tidal wave of hopeful prospectors from all parts of Brazil, starting early in 1988.

Brazil's modern Klondike saw an estimated 45,000 *garimpeiros* (prospectors) flooding into Roraima, illegally, and invading Yanomami lands. This was the new El Dorado.

The gold mining techniques had changed little since the Californians' Forty-niners over a century earlier. Armed with picks, gold pans and high hopes, the *garimpeiros* headed for anywhere it was rumoured that gold had been found. They work manually without heavy machinery, working plots 10 metres square and 1 to 4 metres deep.

Some prospecting is done with a simple pan, but most of the gold is in the form of very fine powder and high pressure hoses are often used to pump mud through machines where the gold can be caught. Gold and impurities are separated by rinsing them in water with mercury. The mercury is burnt off, leaving the precious yellow droplets of gold.

Plate 4 Gold mining at Ouro Verde, Xingu, Brazil. The *garimpeiro* is panning sediment from the valley floor, hoping to find small pieces of gold.

Source: Susan Cunningham/The Environmental Picture Library

Waste water pours into streams and rivers, filling stagnant pools where mosquitoes thrive. The mercury pollution is carried downstream and kills or damages fish and other wildlife.

The work is physically hard, the hours long (miners often work a fourteen-hour day) and the conditions poor. Operations are noisy and muddy, and malaria is widespread. Prospectors abandon their workings when the rainy season arrives, then most return next year.

When all the gold has been extracted, the miner moves on to another plot. The prospectors insist that the forest quickly regrows with no long-term damage, but evidence to support their claim is in short supply.

This massive clandestine gold mining operation is driven by a series of interlocking economic factors. Poverty drove most of the *garimpeiros* to turn their hand to prospecting after Brazil's economic collapse in 1987. Greed allows them to continue operating illegally but with the tacit approval of members of the government and the local establishment, some of whom have financial interests in the operations.

Most of the profits go to mine operators, who take 70 per cent of any gold found. Teams of five *garimpeiros* share the other 30 per cent, from which they pay for their own return flights from the mines and food.

5.5c Problems of prosperity

Traders in local boom towns enjoy their fair share of the spoils, with prices escalated accordingly. The cost of living in Boa Vista, capital of Roraima and centre for the new gold rush, is 40 per cent higher than in Sao Paulo, the country's largest city and business centre. Its airport quickly became the busiest in the continent, with hundreds of light planes ferrying prospectors to the gold workings. It is estimated that up to half of the city's traders turned to gold dealing, and bars and brothels mushroomed in this frontier boom town.

Very little of the gold is marketed legally within Brazil. Much is smuggled out of the country to fetch better prices on the international market. It is widely believed that much of the gold is used in illegal cocaine trading by the infamous Medellin cartel in Colombia.

Gold also buys the silence of some of the authorities responsible for protecting the Indians' rights. It is not uncommon for officials of FUNAI to accept bribes not to protest about the damage and destruction being caused by the prospectors.

Cycles of boom and bust reflect variations in gold prices on the international markets. When the price of gold collapsed (by two-thirds) early in 1990 as a result of the Brazilian government's anti-inflation plan, many thousands of gold miners were abandoned in the Amazon rainforest without food or transport.

5.5d Threats to the Yanomami

The illegal gold mining activities have had a number of serious consequences for the Yanomami. Their tribal lands have been invaded, and there have been incidents of rape and murder committed against the Indians by prospectors. Large areas of rainforest have been cut down to make way for the mining, both at the prospecting sites and to service their needs. The eastern forests in Roraima state are now dotted with over 100 airstrips to which planes and helicopters ferry food, fuel, men and machines. Many camps, where most *garimpeiros* live, radiate out from the bases.

The Yanomami are also threatened by food shortages because miners catch fish and shoot monkeys, wild pigs, *antas* and capybaras, and their noisy operations chase game and fish away from the area. Hungry Yanomami deprived of their natural livelihood are known to beg food from the gold miners, but the processed food brought from the cities causes an imbalance in their diet which leads to malnutrition.

Problems also arise from the pollution of streams and rivers with mercury used in gold panning. This kills fish and poisons humans.

Without doubt the most serious problem is associated with the spread of diseases previously unknown to the isolated Indians and introduced by the prospectors. Malaria, measles, tuberculosis, pneumonia, river blindness and Chagas' disease have reached almost all Yanomami communities with devastating consequences. Malaria is a widespread and fast-growing problem, with many Yanomami areas which were free from malaria in 1987 now up to 90 per cent affected (with estimated death rates from 12 to 68 per cent). Contact with outsiders has also introduced AIDS to the area. Official health statistics show that Roraima has the third highest incidence of AIDS in Brazil (after Rio de Janeiro and Sao Paulo).

Surveys indicate that from 1987 to 1990 the Yanomami dwindled from about 9,000 to 7,500 because of malnutrition, sexually transmitted diseases and malaria. Lesser problems include the spread of skin diseases among the Indians after the introduction of clothing, and widespread tooth decay caused by sugar-rich western food.

A small number of Yanomami have been killed by prospectors in violent clashes. But the miners are mostly friendly to the Yanomami and give them food and clothing. The Yanomami's traditional reputation for ferocious aggressiveness doubtless plays a part, coupled with genuine reluctance among many miners to add to the Indians' problems.

The Brazilian government's reactions to the illegal mining has been slow, partial and partisan.

5.5e Loss of Yanomami lands

The second major threat confronting the Yanomami has been the forfeit of tribal lands by government decree. Their traditional lands have been eroded from an original 95,000 km^2 to 24,000 km^2, and divided into nineteen 'islands' within a larger area designated as a national forest (Figure 5.2).

The Yanomami land has long been in contention. From the early 1960s many groups have been campaigning for the protection of the Indians' constitutional rights, including anthropologists, Survival International, the Catholic Church, Funai and CCPY (the Commission for the Creation of the Yanomami Park, set up in 1978). A major breakthrough came in 1982 when Brazil's Minister of the Interior declared as Indian land a large continuous area covering most of the land traditionally occupied by the Yanomami.

But the plight of the Indians was not looked upon sympathetically by the national government. Under President Sarney direct control of all Indian and environmental policy in the Amazon had been taken over by SADEN (the Secretariat for National Defence, previously the National Security Council). Indirect military control of the Indian lands was justified by the argument that national security and sovereignty are at stake in the border areas, where most Indians live. The Indians were treated as subversive and anti-patriotic under this military regime.

The legislation prepared to confirm the 1982 declaration of Indian land finally appeared in February 1989, signed by President Sarney. It fell a long way short of what the Indians and their supporters wanted, expected or thought fair. Only 30 per cent of the traditional territory of the Yanomami was demarcated as Indian land.

To compound the problem ever further, this land was to be fragmented into Indian colonies (where the Indians would live) and national forests (where economic activities such as logging and mining would be allowed). The 1989 decree split the Yanomami land into a series of nineteen relatively small separate islands within a sea of national forest.

The plan envisaged separating miners from Indians by moving the *garimpeiros* into three areas of national forest, early in 1990. But in October 1989 the plan was declared unconstitutional by a federal court, which ordered the complete removal of the prospectors.

In April 1990 President Collor (the first elected President since Brazil's military take-over in 1964) ordered the federal police to dynamite the illegal airstrips used by the prospectors. The military were slow to respond, and only destroyed about a dozen. The order was later cancelled, and has since been dismissed as a publicity exercise designed to persuade the Brazilian public that the problem was being dealt with.

The operation to drive the gold miners out of Yanomami land collapsed after threats of violence from miners' leaders. FUNAI was ordered to organise the withdrawal of the miners, but – starved of funds by the federal government and seriously under-resourced – it has found itself almost totally incapable of doing so.

Later in 1990 the Brazilian government gave in to the prospectors and allowed them to stay on Indian land. The prospectors' victory was seen as another nail in the coffin of the Yanomami. In late 1991, however, President Collor announced the establishment of a major Yanomami reserve.

5.6 THE KAYAPO OF AMAZONIA

The Kayapo Indians of Amazonia provide another illustrative example of long-term adaptation and the problems posed by recent pressure.[27]

5.6a Traditional lifestyle

Descended from rainforest people who migrated south from Mexico and Central America over 10,000 years ago, the Kayapo were among the very first humans to live in Amazonia. In common with other tribal groups, they gradually evolved a lifestyle suited to the Amazon. They use many different food sources, as hunters, gatherers and shifting cultivators.

Like other forest people elsewhere their way of life is sustainable and has a very limited impact on the forest environment. It is not in their interest to over-exploit the natural forest resources on which they rely, nor to over-cultivate soils or leave inadequate fallow periods for vegetation and soils to recover between cropping cycles (see section 3.3).

Kayapo villages are traditionally sited in ecologically rich transition zones between different vegetation types to give them access to a variety of forest products and soils. They use over 2,500 native rainforest species of plants for food, building, craft and medicine. Women of the tribe look after small intercropped gardens in which they grow up to fifty native plants including fruit trees (such as papaya, mango and avocado) and a range of vegetables, beans, manioc, yams, medicinal plants and natural pest control species. Their small garden plots are periodically abandoned to allow natural regrowth of vegetation and restoration of soil fertility.

This pastoral scene has been much disturbed over the centuries by contact with outsiders, little of it sought or welcomed by the Kayapo themselves. Kayapo territory was first invaded in the sixteenth century by Portuguese gold prospectors. Natives and prospectors fought for more than two hundred years until the Portuguese were able to establish a caravan route through the area to transport gold from Goia and Cuiaba to Sao Paulo.

5.6b Contact with outsiders

Over the nineteenth and twentieth centuries the northern Kayapo had call to resist invasion of their territory on a number of occasions. Close contact with traders and missionaries brought diseases which killed many Kayapo. Indeed, it is estimated that 85 per cent of them died within the first generation of contact with outsiders, from western diseases such as mumps, flu, measles and whooping cough. Villagers are now inoculated against such diseases, but up to a third of the people in a Kayapo village can still die in one epidemic. Other health hazards have been created in recent years by hydroelectric schemes (see section 3.9) which have introduced breeding grounds for insects. Many Kayapo have been killed by diseases like typhoid and malaria.

The surviving Kayapo people now live in a protected area, the Gorotire Reserve in the south-west of Para State in eastern Amazonia (Figure 5.1). The reserve is recognised in law, covers an area of 33,000 km^2 and houses around 1,800 Kayapo living in four settlements.

Although this formal protection has given some welcome relief to the persecuted Kayapo, their future is still far from secure. They are under attack from various quarters. A modern gold rush began in Gorotire in 1983 and within a year around 3,000 prospectors were working sites at Rio Branco, causing heavy mercury pollution along local rivers. By late 1988 there were 4,000 miners working at Maria Bonita near the reserve, again poisoning the rivers with mercury, introducing diseases and seriously disrupting local wildlife.

Since 1981 the edge of the Gorotire Reserve, close to the main settlement, has been exploited by loggers in search of valuable mahogany (although the logging company was ultimately expelled by the Indians).

The Kayapo provide an interesting microcosm of tribal groups in rainforest areas which are being cleared. Their determination to survive has been tested repeatedly as the rainforest which has been their home for thousands of years has been exploited, damaged and cut down. Without government commitment to provide protection for such groups they are caught in an unending spiral of persecution, repression and abuse of their native rights.

5.7 THE PEOPLE FIGHT BACK

Confronted with such pressures for survival, many indigenous groups have fought long campaigns to try to preserve their traditional lifestyles and their native dignity. There is a long history of tribal resistance in the Amazon rainforest.[28]

It is not simply a matter of retaining tribal lands (though that is of fundamental importance). Cultural survival requires protection from introduced diseases, time for threatened peoples to adapt and the right to determine their own future.

5.7a Eviction and resistance

Sometimes the tribal groups take their future into their own hands. In 1976, for example, the Gavioes tribe in Brazil was threatened with resettlement and land appropriation by the government.[29] They resisted by taking control of the collection and marketing of the Brazil-nuts on their land and driving away the unscrupulous government agent who had previously controlled the operation. When power line construction was scheduled to cross their land, the Indians went to court to fight for compensation. They won, and received $830,000. By 1980 the Gavioes were making a profit of about $50,000 a year from their Brazil-nut operation, having also preserved their land, culture and lifestyle.

Another revealing example of tribal people's resistance movements comes from Irian Jaya (West Papua).[30] Here the rainforest people have faced eviction from their tribal lands to allow colonisation by settlers from the Indonesian government's Transmigration Programme (see section 3.5). This ambitious but ill-conceived scheme was designed to move millions of poor families from the overcrowded central islands of the Indonesian archipelago to the relatively undeveloped outer islands. By the end of 1984 some twenty-four major transmigration sites had been established in West Papua and around 700 km^2 of land had been taken from its traditional owners. Many tribal people were forcibly evicted in the process. Opposition and resistance by tribal peoples have been met with attacks by the Indonesian armed forces in what the government called 'Operation Clean Sweep'.

5.7b The Penan of Sarawak

There are many other tales of tribal people fighting back when faced with external pressures. In Sarawak, Borneo, the Penan and other tribal rainforest groups (including the Iban, the Kayan, the Kelabit and the Kenyah) have been engaged in a long and acrimonious struggle against commercial loggers intent on destroying their tribal homelands.[31] The pressures are immense. Three-fifths of the land area of Sarawak is leased to logging companies and logging is claiming more than 2,000 km^2 of forest a year. The tribes depend entirely on the forest for their existence – hunting wild game with hand-made blowpipes, collecting sago from wild palms, collecting herbs, medicines, waxes and resins.

The Penan faced a similar problem to many rainforest tribes, in that the government allocated logging concessions unilaterally, without seeking or being granted any consent from the tribal groups whose homelands were being given away. Appeals to state and federal authorities to have their traditional land rights formally recognised have gone unheeded.

Plate 5 Penan tribesmen blocking a logging road at Umbang, Sarawak.

Source: Nigel Dickinson/The Environmental Picture Library

Faced with a government indifferent to their claims, the Penan had few option other than direct action against the loggers who were removing their heritage tree by tree. A well co-ordinated series of fifteen blockades of the logging roads and tracks leading into their lands was mounted in April 1987 (just before state elections), involving hundreds of men, women and children from different tribal groups.[32] It was very effective, and lasted for eight months. Meanwhile the logging activities were just about brought to a standstill, because neither loggers nor logs could get through the human blockade. This classic David-and-Goliath situation was eventually ended in October 1987 when armed police moved in to dismantle the blockades. Many tribesmen were injured and forty-eight were arrested.

The government subsequently introduced tough legislation to ensure that the rights of commercial loggers were protected at the expense of the rights of tribal peoples. Under the new legislation, anyone found guilty of interfering with logging activities in Sarawak would be liable to a $6,000 fine and up to two years' imprisonment.

The resolve of the Penan and their fellow tribespeople was not to be broken, however, and they have continued to use the defiant and effective tactic of forming human blockades to disrupt logging. Between November 1988 and January 1989, 128 people were arrested under the new law.[33] The loggers appear to be winning this particular battle, because by 1989 logging was proceeding round the clock, using spotlights at night.[34]

Sometimes the survival campaigns have been fought by the native tribes alone. Many tribes and peoples are being threatened by similar pressures, so there is strength to be gained from working together and pooling resources and people-power. The various Amazon tribes, for example, have set up the Union of Indian Nations to defend their land and rights. The aims of the union include 'promoting cultural autonomy . . . fighting for land and resource rights, helping communities in designing and carrying out development projects, and informing the general public about indigenous peoples'.[35]

Well co-ordinated and publicised public protests stopped the plan to flood parts of Tasmania's rainforest at Lake Pedder by the proposed Franklin Dam. Public interest around the world was heightened by the syndication of film coverage of the world-renowned ecologist David Bellamy protesting and being arrested.[36]

5.7c Rainforest martyrs

The stakes are high and the struggle is often acrimonious if not hazardous. People who defend tribal territories and human rights are often persecuted and punished. The silence of some opponents to forest clearance has been bought by their murder. The struggle to protect the forests is not without its heroes and martyrs. Over a thousand rubber tappers, peasants, Indians and union officials have been killed in South America since 1980.[37] Vicente Canas, a Jesuit, was murdered while attempting to protect a Brazilian Indian tribe in the Mato Grosso.

The most prominent rainforest martyr is Francisco Alves Mendes Filho, better known around the world as Chico Mendes. Mendes founded the Brazilian Rubber Tappers' Union and lobbied tirelessly for the rights of all people whose livelihood depended on the survival of the forests. He quickly became champion of the rainforest cause, and in 1988 he wrote: 'I believe that in a few years the Amazon can become an economically viable region not only for us, but for the nation, for all humanity, and for the whole planet.'[38] A few months later, in December 1988, he was assassinated by gunmen hired by the Uniao Democratica Rural (UDR), a vigilante association of ranchers and landowners.[39] He was the fifth rural union president murdered in Brazil that year.

5.8 THE ALTAMIRA GATHERING (1989)

Media attention is a potent factor in raising awareness around the world of the plight of the threatened rainforest tribes. One recent event which captured widespread television and newspaper interest was a tribal gathering near the town of Altamira in Brazil (Figure 5.1), in February 1989. The two-day meeting was held at one of the sites planned for a major hydroelectric power development (see section 3.9).

It was the largest ever gathering of indigenous peoples, attended by an estimated 3,500 warriors from twenty-eight Indian nations.[40] Amongst the tribes taking part were many who have been seriously threatened by development programmes in Amazonia, particularly the dam proposals. Representatives of the Parakana Indians, who were resettled when their tribal lands were flooded by the Tucurui Dam (see section 3.9), were there. So too were members of the Waimiri-Atroari Indians, who were affected by construction of the Balbina Dam (see page 77). Other tribes represented include the Xavante, the Arara, the Gaviao and the Yanomami (Figure 5.1).

One aim of the gathering was to establish a permanent settlement at Altamira which would act as a focal point for all the Indian tribes represented. The primary aim was to provide a focus for bankers, politicians and other interested parties to discuss what was going on in Amazonia, and to exchange views about priorities and about what the future held in store.

The organisers of the gathering took every opportunity to turn it into a media event, and the scene was recorded by 200 of the world's press and over 100 international observers.

Plate 6 The Altamira Tribal Gathering, Brazil. Forest people from Brazil were joined by warriors from 28 Indian nations at the tribal gathering near Altamira, in February 1989.

Source: Susan Cunningham/The Environmental Picture Library

The gathering was remarkably effective in focusing attention of the real problems and costs of rainforest clearance.[41] One direct consequence was to the withdrawal of Power Sector and development loans worth US$1,100 million from the World Bank and other agencies. The Brazilian government quickly set about creating a series of environmental programmes, costing between US$300 and $400 million.

5.9 CONCLUSIONS

The indigenous groups represent all that is good about the rainforests. Their fate is intimately tied up with the fate of the forest. Both are under attack, and both represent resources the world can ill afford to loose. Yet to argue that these groups can and indeed *should* be totally and permanently shielded from acculturisation and contact with outsiders is clearly nonsense. What is required is an acceptable rate of change in their style of life, and change based on the choices of the indigenous people themselves.

As Catherine Caufield stresses, at risk is 'not only the skills but also the wisdom, the social patterns, and the outlook of aboriginal peoples [which] constitute a fund of knowledge that, like a genetic bank of wild plants, technological man may need to call upon in the future'.[42]

6

POSSIBLE SOLUTIONS

For the purposs of recreation he has selected the felling of trees, and we may usefully remark that his amusements, like his politics, are essentially destructive.

Lord Randolph Churchill, on Gladstone, 1884

6.1 INTRODUCTION

We have seen in earlier chapters that rainforests are the most diverse and productive ecosystems in the world (Chapter 1), but they are also the most threatened (Chapters 2 and 3). Deforestation removes much more than just trees, and clearance of the rainforest causes wholesale extinction of species and the removal of important environmental services, and promotes climatic change (Chapter 4). Forest peoples suffer, too, and whole tribes and their cultures and knowledge – accumulated over generations of living sustainably in the forest – are being lost (Chapter 5).

Clearly this state of affairs cannot be allowed to continue unchecked; the stakes are simply too high, and it is a global not just a national problem. In this chapter we examine what options are available to solve the problem, and evaluate recent national and international initiatives designed to protect the remaining rainforests.

6.2 THE NEED FOR ACTION

Whilst different experts and agencies prefer different solutions (Table 6.1), there is widespread agreement that action is required. The consensus also recognises that it is required urgently, because the very survival of the rainforest is under serious threat. Recent estimates[1] show that, if present rates of clearance continue, all remaining rainforests are likely to have been destroyed by the year 2000 (with the possible exceptions of small areas in western Amazonia and central Zaïre; see section 2.5).

Table 6.1 Solving the problem of rainforest clearance: solutions proposed by the *Global 2000 Report*

Proposed solutions include:

*better management of existing forest resources
*reforestation
*tree plantations
*rangeland management with controls and pasture improvement
*restriction on new land clearing (based on soil capability studies)
*development of agro-forestry techniques for people who now have no alternative to planting annual crops on steep slopes
*dissemination of more efficient wood cooking stoves
*development of biogas and solar stoves to replace wood and charcoal burners
*intensification of agriculture and other employment-creating forms of rural development, in order to reduce agricultural pressures on remaining forest lands

Source: summarised from Barney (1980).

Many international conservation agencies (such as the World Wide Fund for Nature and Friends of the Earth) have designated the saving of the rainforests as their single most important conservation priority.

Prince Charles captured the feelings of many who are concerned about the forests, in a special Rainforest Lecture he delivered at Kew Gardens in London in 1990. He stressed[2] that

> we are literally the last generation which can save the rainforests from total destruction . . . if we don't act now, there won't be much rainforest for our children to be concerned about. . . . For hundreds of years, the industrialised nations of the world have exploited – some would say plundered – the tropical forests for their natural wealth. The time has come to put something back, quickly.

Added urgency enters the argument when new statistics on rates of forest clearance are published. The trend is largely one-way – upwards. A Friends of the Earth report by Norman Myers, published in 1989,[3] claimed that the rate of tropical deforestation had increased by 90 per cent during the 1980s. Myers estimated that 142,000 km^2 of rainforest were cleared in 1989 alone. An estimated 170,000 km^2 were cleared in 1990, according to an August 1991 report by the UN Food and Agriculture Organisation which found the greatest rates of clearance in Latin America, followed by Africa and Asia.

There is no shortage of technical solutions to the rainforest problem and many experts insist that what is really missing is a broader awareness of the problem and a commitment to act. But it is not quite so simple as that, because it is also widely agreed that 'there is no single action which would halt or even significantly slow the rate of tropical deforestation'.[4]

6.2a The case for protection

It should, by now, be largely self-evident why action is needed to save the rainforests, rather than simply letting market forces prevail and sit back while those that remain are cleared.

We value the forests in a variety of ways, both extrinsic and intrinsic (see section 1.9). The environmental and climatic services performed by the forests (Chapter 4) are highly important. Forest peoples have a right to survive, along with their lifestyles, culture and indigenous knowledge (Chapter 5).

The case of protecting remaining rainforests has been put many times and in many different ways. Over two decades ago, the vegetation geographer Robert Eyre concluded[5] that 'it would be tragic for both science and the economies of the tropical countries if such a large and varied resource was simply allowed to dwindle away'.

Prince Philip, a lifelong and world-famous conservationist, has highlighted some of the complexities of the debate. He has noted[6] how

> to the eyes of conservationists the great natural rainforests are objects of beauty and delight and the habitat of a vast range of plant and animal life. They are also vital components of the world's climatic conditions and the carbon dioxide/oxygen cycle. But to the governments of these countries, struggling with internal hunger, poverty and unemployment, with external debt and with a plethora of well-meaning advisers and foreign aid programmes, the forests represent wealth in a fairly easily convertible form.

This persistent tension between regarding rainforests as a capital asset there to be exploited, or as a non-renewable natural resource of global significance which must be protected at all costs, lies at the heart of the debate. Yet the forces of destruction march relentlessly on.

A recent review of the rainforests[7] concluded that

> the arguments for halting the deforestation of the rainforest range from aesthetics, through scientific curiosity, scientifically based resource preservation, economically motivated control of supply to humanitarianism. On present showings, none of these is likely to curb rainforest destruction before more of the damage is done.

Herein lies the ultimate irony surrounding the rainforest debate.

6.2b Activism and demonstrations

A key issue in mobilising public awareness of the problems of forest clearance, in both tropical and developed countries, has been direct action. Sometimes this has involved tribal peoples in demonstrations against intruders, like the blockades of loggers' access roads into the Sarawak forests by the Penan[8] (see section 5.7).

More media attention is usually won by specially staged events such as the demonstration to prevent flooding of Tasmanian rainforest by the proposed Franklin Dam. The 1989 gathering of Brazil's forest Indian tribes at Altamira (see section 5.8) was particularly effective in attracting the attention of the world press and, through that, millions of people around the world. In July 1991 eight foreign environmentalists (from the UK, USA, Germany, Sweden and Australia) were arrested by the Malaysian authorities after staging anti-logging protests on the island of Sarawak.

Yet not all demonstrations are large or acrimonious. Amongst the most poignant (and photogenic) displays of concern about the plight of forests and forest people are the tree-hugging pursuits of the Chipko Andolan movement of the Indian Himalayas.[9]

6.3 CONSTRAINTS

The problem of protecting rainforests is multi-dimensional because the forces of clearance are interdependent and different forests are threatened with different pressures. The task of stopping forest clearance is a tough one, particularly because it requires some (even partial) resolution of complex issues like international debt repayment, rising population and meeting the needs of the landless poor.

6.3a Complications

There are also many thorny questions concerning how to resolve conflicting interests at different scales. Preservation of the rainforest is a global priority, but deforestation is promotedby certain groups either within a country (such as the gold miners in Brazil; see section 5.5) or internationally (such as the demand for tropical hardwood products in developed countries; see section 3.6).

Jimmy Carter, then President of the United States, was advised in 1980[10] that

> the tragedy of the forests is that, like the commons, they are . . . subject to misuse – but on a global scale. While forest lands are owned by individuals (or governments), forests provide community, national and international benefits that go well beyond the benefits usually considered in forest management decisions [and] . . . do not enter into the normal calculus of forestry economics.

Added complication comes from the fact that even if forest clearance was to be stopped entirely, it is doubtful whether many badly disturbed rainforest ecosystems could restore themselves by natural regeneration. Permanent damage has already been done in many areas of clearance. Species are lost, either to an area or by total extinction (see section 4.2), and soil erosion and loss of fertility severely inhibit the prospects of self-repair (see section 4.4). Moreover, the time-scale required for even partial recovery would be immense.

Forest replanting and restoration schemes might have potential in some areas, but the costs are likely to be prohibitive. Even if it were possible, the prospects of restoring much of the inherent diversity of natural rainforest are limited.

Any realistic attack on the problem of tropical deforestation, therefore, must be based on a broad strategy in which different pressures are tackled in appropriate ways. The solution ultimately lies in better international awareness and commitment to act. After all, rainforest clearance means significant extinctions of species and threatens to trigger climatic changes which could ultimately affect most parts of the globe.

6.3b Research, training and education

Most experts and agencies involved in forest management recognise the vital part played by research and education. Without a better understanding of how the forest works, what influences its structures and dynamics, and how resilient it is, there is little prospect of any sustainable future for those forests which remain. Without a well-conceived and broadly disseminated programme of education about the forests, their uses and potential, it will be difficult to change public opinion or remove some of the pressures facing the forests.

There have been numerous calls for major investment of time, money and trained personnel into research on rainforest ecosystems.[11] While scientific research in recent decades has revealed a great deal about this unique and highly complicated ecosystem (see Chapter 1), important questions remain unanswered.

There are still many gaps in our understanding of the structure and ecology of rainforests, for example, as well as unknown numbers of species of plants and animals which have yet to be discovered, classified and named (see sections 1.8 and 4.2). Whilst forest people have evolved sophisticated methods of using rainforest products sustainably (see section 5.2), modern scientists and foresters have many unresolved mysteries about how to achieve the sustainable development of suitable areas.

There have also been calls for improved education programmes, directed at local populations and people in developed countries who reap the benefit of tropical deforestation. Wholesale changes in attitudes towards the rainforests are called for, away from exploitation towards conservation.

6.3c Land reform

Some of the pressures on the rainforests – such as the recent rise of shifting cultivation (see section 3.3) and the impacts of population transmigration schemes (see section 3.5) – arise from land hunger promoted by unequal ownership.

In the Latin American countries as a whole, for example, about 93 per cent of the arable land is owned by 7 per cent of the landowners.[12] In Brazil, 1 per cent of the farms occupy over 43 per cent of the total farmland, half of the farms are squeezed onto less than 3 per cent of the farmland, and 7 million families are landless.[13] In Java over half of the landowners possess less than half a hectare (5,000 m^2) of land each, while 1 per cent own about a third of all the land.[14]

The net effect of such inequalities in landownership is mounting pressure on the landless peasants to claim land for themselves. This is normally easiest to do by simply clearing a patch of forest, laying claim to it and farming it. But this poverty trap is placing growing pressures on many forest remnants, especially those which have recently been opened up by new roads or logging tracks.

The only way to remove this sort of pressure on the rainforests is to tackle the landownership question by wholesale land reform. Some critics of the Brazilian government's policies towards land stress the need to remove responsibility for the fate of the forests from the hands of a small rich elite (the large estate owners and outside agencies whose main interest is rate of return on capital investment), and place it back in the hands of those to whom it rightly belongs (the people of the country).

Although many experts agree that land reform is desperately needed, the evidence from many countries is not promising. In Brazil, for example, forest peoples have had much of their tribal territories confiscated by government decree (see section 5.4). Even though land is dedicated to the tribal people in Indian reserves, the Indians do not own it. The Brazilian government still retains control of the title on that land. Peru and Venezuela have also confirmed Indian rights, but land given to the Indians is too small to support their lifestyle of migrant hunting and cultivation.

Some countries have been more positive. The Colombian government, for example, has given 178,000 km^2 of Amazon rainforest to its native Indian peoples, helped by a European Community grant of $386,000 to help administer the system.[15] Native Indians administer the land and control who may enter it, but the government reserves the rights of mineral and commercial extraction.

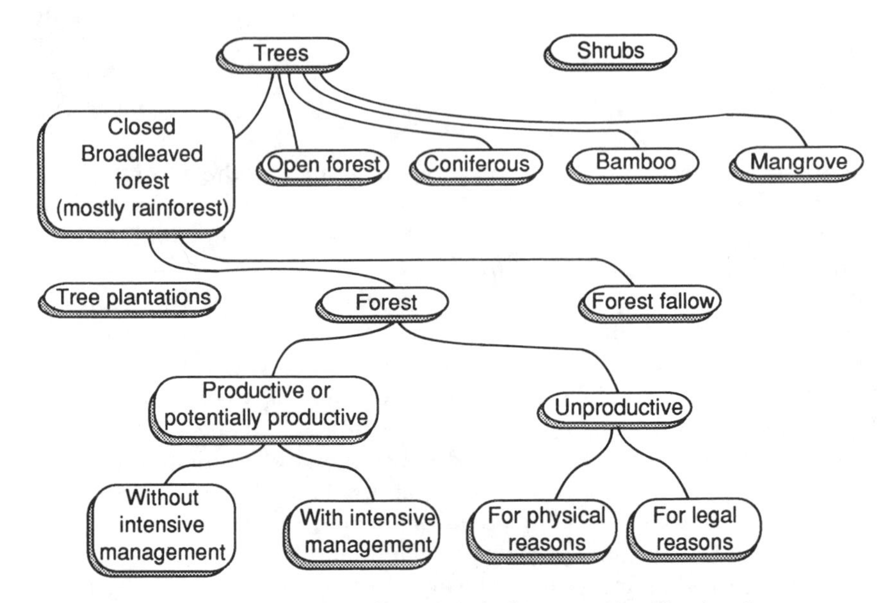

Figure 6.1 The United Nations Tropical Forest Classification System.

Source: after Johnson *et al.* (1989b)

6.4 PROTECTION AND CONSERVATION

A common and effective approach to protecting nature in many countries has been to designate particular areas as national parks or nature reserves, and restrict land use changes or damaging activities within the designated areas. Not surprisingly, there have been many calls for the establishment of rainforest parks and reserves (which are classed as 'unproductive' for legal reasons in the United Nations Tropical Forest Classification; Figure 6.1).

6.4a Historical context

The first concerted international efforts to slow down deforestation around the world came in 1973–4, in the Man and Biosphere (MAB) programme organised by UNESCO.[16] MAB sponsored and co-ordinated scientific research in many of the world's threatened ecosystems, including rainforests.

Proposals from the MAB studies include the suggestion that all governments should lay aside tracts of land in which development would be prohibited. The programme also recommended that governments fund research into the functions of the rainforests so that ecologically sound management plans could be devised. As often happens, however, governments took little notice of the MAB recommendations and the programme has had little practical impact on the fate of the rainforests.

Setting aside protected areas was also favoured in the World Conservation Strategy (WCS)[17] formulated by the World Wildlife Fund (WWF, now the World Wide Fund for Nature) and the International Union for the Conservation of Nature (IUCN). The WCS proposed a planning framework which some countries (like Zambia and Madagascar) have used to establish forest conservation schemes.

A limited amount of financial assistance to help establish protected areas of global importance has been made available under the World Heritage Convention,[18] although by the start of the 1990s very few tropical forests had been included in the scheme.

Public involvement in rainforest conservation has been encouraged by the Fondation Amazonie, a controversial new Amazonian rainforest group set up in September 1991 by Belgian film maker Jean-Pierre Dutilleux (who was previously associated with Sting's Rainforest Foundation; see section 5.1). Donors are invited to help purchase strips of rainforest to enable the creation of a national park, three times the size of Belgium, in the upper Solimnoes River region of western Amazonia, at a total cost of £3 million.

6.4b Ecosystem conservation

It is imperative to conserve natural ecosystems intact rather than just individual species, for various reasons. The environmental services performed by rainforests (see section 4.4) can only be protected in working ecosystems. Many species have co-evolved, which means that they have evolved to rely on each other and can only be conserved in their natural habitat.

An alternative to conserving ecosystems might be to extend the rescue and reintroduction programmes operated by zoos and botanical gardens, but this would only benefit some species. It is estimated that all the zoos in the world could only maintain viable populations of around 900 animal species; this would leave many rainforest species without protection.[19]

6.4c Designation and protection

Some areas of rainforest are already protected as National Parks and nature reserves. In 1990 there were roughly 560 tropical forest parks and reserves covering a total of 780,000 km^2 and accounting for about 4 per cent of all tropical forests.[20] The total area protected as reserves is small compared with the area which is logged or otherwise put to productive use (Figure 6.2).

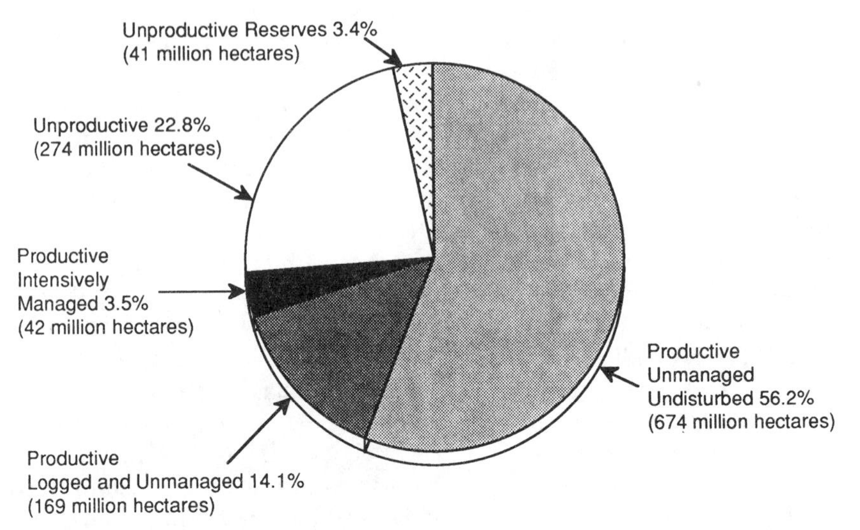

Figure 6.2 Management of tropical closed forests, 1987.

Source: after World Resources Institute (1987)

There have been some remarkable success stories which illustrate the potential of this protection approach.[21] The establishment of the Parc National des Volcans in Rwanda, one of Africa's smallest but most densely populated countries, for example, has allowed the endangered mountain gorilla and its habitat to be saved from extinction. Many mammals and some of the eighty endemic Sulawesian birds restricted to primary rainforest live in the 3,000 km^2 Dumogo-Bone National Park in Sulawesi (Indonesia). Over half of the remaining rainforest in Nigeria is to be protected in the proposed Cross River National Park, which is likely to encorporate the adjacent species-rich Boshi-Okwango Forest Reserve.[22]

The Sao Paulo state government has declared 10,000 km^2 of Atlantic forest in Brazil an environmental protection zone. This is a particularly important area of rainforest because many of its species are endemic. Around 80 per cent of the primates, 39 per cent of the mammals and 54 per cent of the birds of the Atlantic forest occur nowhere else in the world.

In October 1990 the European Community agreed massive financial support for a project to protect tropical rainforest in Central Africa. The grant of 24 million Ecu is intended to establish a regional network of protected areas in seven countries which will be used for research purposes to help protect rainforests elsewhere. The scheme will allow local people to continue to exploit the forest in sustainable ways (see section 6.6).

6.4d Korup rainforest project, Cameroon

An interesting example of the difficulties of establishing protected areas is the initiative being taken by the government of the United Republic of Cameroon in Africa.[23]

Cameroon houses some of the last extensive tracts of rainforest in Africa, particularly in the forests of Dja, Pangar-Djerem and Korup. But these forests are under immense pressure in a country where four out of five people are engaged in agriculture and the population is expected to double between 1980 and 2020. The Cameroon government is committed to conservation, and its conservation objectives are being integrated with the country's economic development plans at national and regional levels.

One conservation objective is to establish a network of National Rainforest Parks covering a total area of 11,000 km^2 and representing a tenth of the world's effectively protected tropical forest. The network would include the richest tropical rainforests left in Africa. The first park, covering an area of 1,250 km^2, was established in 1986 at Korup in South-West Province, near the border with Nigeria.

Preliminary surveys carried out in Korup in the mid-1980s revealed some of the ecological potential of the rainforest. Seventeen tree species previously unknown to science were discovered, and Korup emerged as botanically the richest and most diverse forest yet studied in the whole of Africa. An intensive chemical screening programme identified more than 90 substances of potential economic value, 38 of them new to science.

The Korup and related projects require massive financial and technical assistance, from the Cameroon government and elsewhere. In April 1988 the WWF pledged to raise £2.5 million, Britain's Overseas Development Administration (ODA) pledged £500,000, and the West German government pledged £600,000 to the project. WWF signed a contract with the Cameroon government to establish a protected core area of 1,200 km^2 surrounded by a buffer zone of 4,000 km^2. The plan included helping farmers in the buffer zone to identify the better land, and offering them advice on farming techniques. Tree nurseries with fast-growing trees would provide timber for everyday needs. The Cameroon government agreed to establish guard posts along the border of the buffer zone to try to stop illegal hunting.

The threat to the Korup forest continues because the Cameroon government sees the country's extensive rainforests as a prize source of foreign exchange to help alleviate its severe economic crisis.[24] Illegal logging continues in the area, in the absence of effective surveillance and policing mechanisms. Logging is even occurring under licence since the government of Cameroon granted logging concessions in areas bordering the National Park to international timber companies in February 1990. Observers report that some logging companies have been taking species other than those they have declared, and that directives to consult local agricultural officials and communities before applying to cut timber are being ignored.

The aim of the Korup project was to protect the rainforest and its 8,000 species of plants, animals and insects, while providing the local population with space for sustainable rural development in a buffer zone around the forest. Despite its problems, the project has been widely regarded as a model for the conservation of threatened rainforests around the world.

6.4e Problems

There are some serious problems surrounding the existing global network of protected rainforests. There are too few sites, for a start. It is almost inevitable that many rare species will become extinct unless more rainforest reserves are set up and protected.

The distribution is far from adequate and some distinct forest types are not represented in the network. Missing pieces in the mosaic include the Malagasy thorn forest, the Sri Lankan and Burman rainforests and the Chilean araucaria forest.[25] Moreover, most of the areas have been disturbed previously so the network does not contain the most valuable natural rainforests.

Existing reserves and parks need much better policing and on-the-ground surveillance. Many designated reserves exist on paper alone, and are not supported on the ground through lack of financial resources.[26] The World Resources Institute points out[27] that 'in the developing world, many of these legally protected and reserved lands are still threatened by poaching, illegal timber harvesting, wildfires and encroachment'.

Demarcation of a protected area does not in itself guarantee protection, and examples of continued destruction of designated areas are legion. For example, in late 1991 logging was continuing in protected areas on the Indonesian island of Siberut, in violation of an international agreement to preserve the area. Plans were by then well advanced to log 1,500 km^2 of virgin rainforest and replace it with palm oil plantation, much of which will be worked by migrant labour from Indonesia's main islands.

6.4f Optimum size of reserve

Many of the individual units are too small to be effective. Conserved forests need to be large and undisturbed enough to be self-perpetuating.[28]

It is difficult to determine exactly what is the optimum or even the viable minimum size for a rainforest reserve, because this depends partly on the species present. But it is generally agreed to be much bigger than most existing parks and reserves, which are between 1 and 25 km^2.

Large areas are essential because many forest species travel over wide areas in search for particular food or habitat (such as a particular forest tree, which may be widely scattered). Apes need particularly large home ranges. Most forest species are reluctant to cross even small areas of cleared land, so the protected area needs to be large if they are to survive.

A major experiment is under way in the Brazilian rainforest to determine what size of units should be conserved. The 'Minimal Critical Size Project' funded by the World Wide Fund for Nature[29] will collect field observations of species distributions and lifestyles over twenty years and use them to estimate the smallest viable size of forest for designation and protection.

The answer to the question 'How much rainforest should be saved?' depends largely on what the main priority is. Preservation of biological diversity is possible within a relatively small total area of rainforest, whereas preserving cultures and lifestyles of indigenous peoples requires a much larger area. Preservation of major environmental functions requires a much bigger area again.

The major drawback with the protection policies is that they fail to tackle the root causes of deforestation. The World Wide Fund for Nature rightly points out[30] that 'conservation groups can never hope to buy all the threatened tropical forests. Nor will governments simply turn them into protected areas with no economic function. Any solution must therefore tackle the economic causes of forest loss.'

Prevention is much better than cure, and there is an urgent need to tackle the underlying causes of deforestation. We return to this theme later (see sections 6.8 and 6.9).

6.5 RESTORATION AND REFORESTATION

As well as setting aside areas of rainforest as National Parks and nature reserves, there is a pressing need to restore badly damaged or degraded rainforests. This is unlikely ever to recreate natural conditions fully, but it is an important means of preserving at least some of the diversity and complexity of natural forests.

6.5a Restoration

Restoration seeks to remove the pressures which are altering or removing the rainforest, and then deliberately manage the forest that remains in such a way as to encourage natural self-repair by the ecosystem. Ecologists[31] recognise a number of priorities in such schemes, including the rapid restoration of the canopy and the characteristic height of the rainforest. Hunting and cutting must be prevented in the forest under restoration, and plants and animals previously known to exist there must be carefully reintroduced.

Whilst rainforest restoration is possible in theory, most forests which have been damaged are so badly degraded that it is ineffective in practice.

6.5b Reforestation

Reforestation involves replanting a cleared area with trees, and then managing it to ensure that the vegetation thrives. If restoration of badly damaged forests has limited potential to recreate natural conditions, the prospects for reforestation are even more restricted. But tree cover, even if it comprises different species to the natural forest it replaces, could still help to restore the environmental services (see section 4.4) and minimise the possible impacts on climatic change (see sections 4.5 and 4.6).

Reforestation is required on a massive scale if rainforest losses are to be made up. Tropical developing countries (excluding China) were planting in the order of 11,000 km^2 a year in the mid-1980s. But it is estimated[32] that replanting needs to be increased by between 10 and 20 times to offset forest losses and meet increasing demand for forest products.

On a smaller scale, local replanting programmes have proved very beneficial in many areas. Some benefits are economic. One commune in tropical southern China, for example, with 10,000 people, planted 3.4 million trees in one year and dug contour ditches to prevent soil erosion.[33] Sale of forest produce, coupled with fish from their fish-farm and electricity from their small hydroelectric generator, has produced an income for the commune which is twice the provincial average.

Other benefits are environmental. For example, reforestation can help to reduce flooding and silting resulting from rainforest clearance. An extensive planting programme in India's Damodar Valley has reduced floods in catchments above dams, decreased sedimentation and increased water supply for agriculture and drinking.[34]

Reforestation is not widely seen as a serious option in tropical areas because of the inherently low soil fertility which might have been made even worse as a result of deforestation (see section 1.6). The laws which permit commercial logging of rainforests in Malaysia also require replanting after logging activities are finished, but few logging companies comply.[35]

Where replanting does occur it is usually commercial plantations with fast-growing trees (used, for example, for pulpwood). But these monoculture stands have their own drawbacks, although they are an improvement on desertified or laterised bare soils.

6.6 SUSTAINABLE USE AND MANAGEMENT

6.6a Conservation versus sustainable development

A fundamental question in the rainforest debate[36] is whether to conserve an area or allow sustainable development within it.

Conservation is normally based on preserving the natural habitat and its indigenous peoples, with no development allowed. The Naional Park and forest reserve schemes are typical. Sustainable development, on the other hand, involves protecting the habitat and people but also allowing a type and level of development that can be sustained into the future with minimum damage to people or forest.

There is mounting support for sustainable development of at least some rainforests, provided that protection is offered as well. This would generate capital which could be reinvested elsewhere in the economy of the tropical countries concerned. Sustainable development would also provide a better way of life for many forest people and allow future generations to share in the benefits of their rainforest inheritance.

A number of strategies of sustainable development of the rainforests are available including sustainable forestry, marketing of non-tree forest products and farming within forests.

6.6b Sustainable forestry

Sustainable forestry usually involves selective logging, in which the commercially valuable trees are removed but the rest of the forest ecosystem is left as intact as possible. However, logging practices generally make such selective action very difficult to guarantee (see section 3.6).

The International Tropical Timber Organisation (ITTO; see section 6.8) opposes non-sustainable logging, and estimates that sustainable tropical timber operations cover only an eighth of 1 per cent of rainforest land.[37]

Many tropical countries whose rainforests are being logged are eager to promote selective logging rather than wholesale clearance. Parts of peninsular Malaysia, Trinidad and Tobago now have selective logging operations, and Costa Rica and Cameroon are exploring the possibility.[38] But sustainable development has yet to be introduced widely into Amazonia, which could really benefit from it, despite commitments to introduce it from all eight Amazon states in 1977.

Countries which import tropical timber are also starting to recognise the need for selective logging. In 1989, for example, West Germany agreed that it would only import logs from sustainable sources.[39]

It is estimated that a permanent forest estate of around 40,000 km^2 is required to satisfy the existing demand for tropical timber within the UK, which is more than four times the area now operated in the tropics under strict conditions of sustained yield management.[40]

Interesting coalitions of groups concerned with the protection and preservation of rainforests have started to emerge. For example, the Timber Trade Federation (TTF) and World Wide Fund for Nature (WWF) issued a joint statement in March 1991 committing both organisations to sustainable and environmentally sound management of the world's forests. The WWF set a target date of 1995 for 'sustainable timber production' world-wide.

Despite the obvious attractions of sustainable forestry practices, some caution has been demanded by the World Bank and other leading funding agencies. They doubt whether sustainability can actually be achieved in practice, and suggest that a sudden halt to commercial logging would result in a desperation phase in which companies would seek to recoup their investments as quickly as possible.[41] If that happened, the transition from clear-cutting towards selective logging could trigger massive forest clearance and wholesale environmental damage.

6.6c Non-tree forest products

Forest people have in many cases been custodians of the forests for thousands of years. Their traditional practices are sustainable (see section 5.2) and it has never been in their interests to over-exploit the forest. So long as their land tenure is secure, it is argued, they are quite capable of producing a surplus of forest products for sale to the outside world.

Some see the involvement of local people as the key to successful management of rainforests. Sustainability will be much easier to achieve if the accumulated knowledge, experience and sensitivity of indigenous people are tapped and forest management schemes are kept small scale and local.

The World Wide Fund for Nature is evaluating a number of promising schemes which would enlist the help of forest people and in return help them develop sustainable income-earning activities (for example, the harvesting of mango trees, freshwater fish-farming and the location of commercially valuable medicinal products in the forest).[42] The aim is to strengthen forest people economically, help to halt the decline of the forest and demonstrate that the forests can be sustainable resources if used properly.

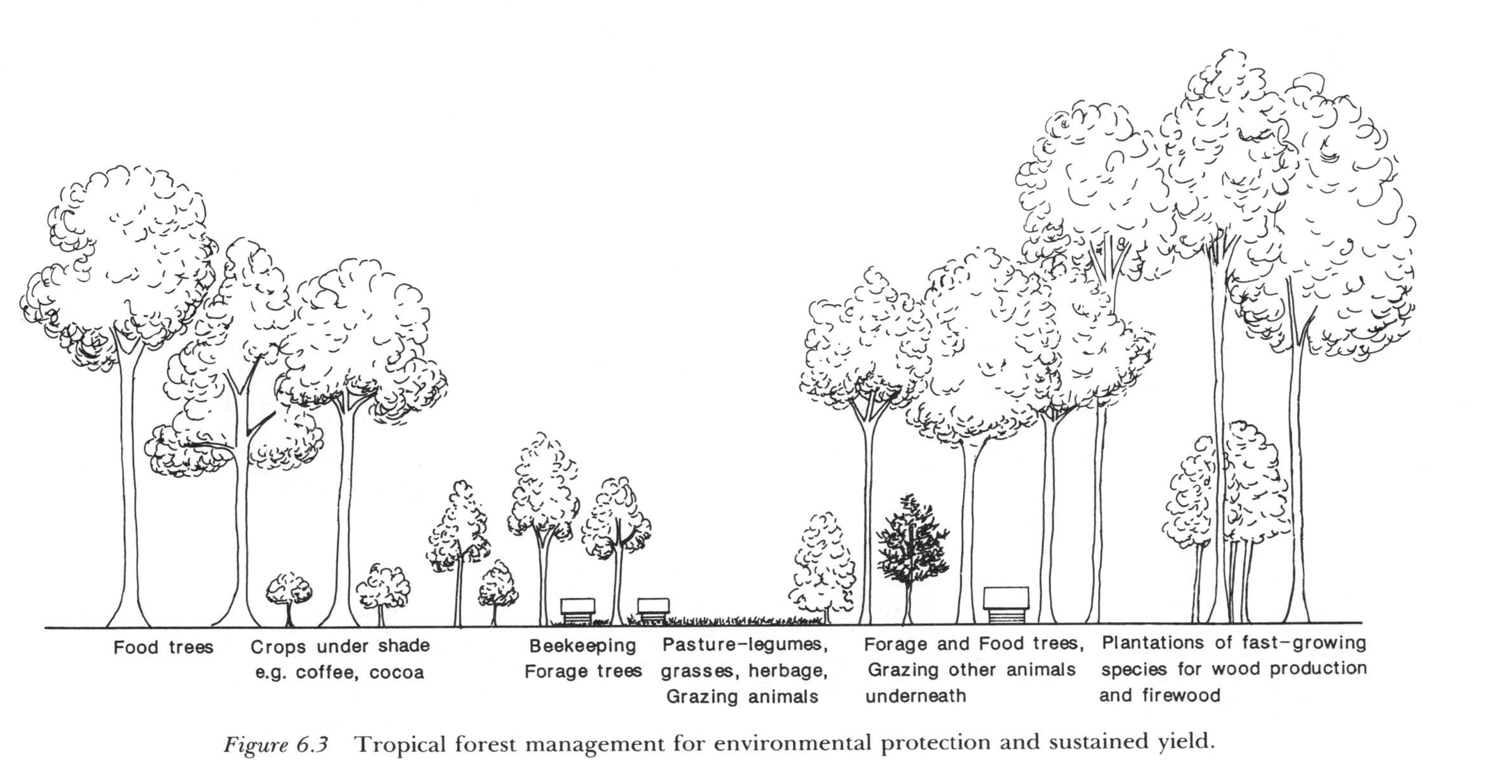

Figure 6.3 Tropical forest management for environmental protection and sustained yield.

Source: after Simmons (1989)

Sustainable forest-based activities do not involve felling of trees and, if carefully managed, they bring little harm to the forest. There are many possibilities (Figure 6.3), but the more common suggestions include rubber tapping, searching for forest plants and animals for use in industry, agriculture and medicine, and collecting vines and rattan for fibre.

Recent success stories[43] include the establishment of a collective by Brazilian rubber tappers to defend 'extractive reserves' and organise the market, and the setting-up of a butterfly farm in Indonesia, to allow local people to sell insects on the European butterfly house market.

In June 1988 ITTO (see section 6.8) approved US$1 million funding for the first phase of a commercial-scale test of sustainable forestry in Acre in western Amazonia. The aims of the project include surveying the economic value of non-timber products, assessing the effects of logging on sustainable productivity and testing the viability of 'extractive reserves' (areas open to sustainable development, including harvesting rubber and collecting Brazil-nuts) within the forest.

The World Bank and WWF launched a pilot project in Madagascar in August 1988 aimed at halving forest clearance there within two decades. The viability of a range of schemes is being evaluated, including fish-farming, water harvesting, terracing to prevent erosion, agro-forestry and fuelwood plantations.

Switching to developing non-timber products not only means less pressure on the surviving rainforests; it can also make good business sense. A paradox of tropical deforestation is that the non-timber products which are literally burned or thrown away are often more profitable than the timber.

A recent study of an area of rainforest along the Rio Nanay in Peru[44] found that 'fruits and latex represent more than 90 per cent of the total market value of the forest, and the relative importance of non-wood products would increase even further if it were possible to include the revenues generated by the sale of medicinal plants'. The price tag put on forest products was very revealing; timber value per hectare was about US$1,000, compared with a sustainable revenue of about US$700 per annum for renewable non-tree resources. The study concluded that 'without question, the sustainable exploitation of non-wood forest resources represents the most immediate and profitable method for integrating the use and conservation of Amazonian forests'.

6.6d Farming within the forest

An alternative means of making sustainable use of the rainforests and at the same time helping forest peoples is to encourage small-scale farming in plots within the forest (agro-forestry). This promotes continuity of traditional farming practices, prevents wholesale clearance of the forest and keeps the future of the forest safely in the hands of those who know best how to use it and protect it at the same time.

The difficulties of co-ordinating and marketing such activities is perhaps best done at the local level, by community co-operative schemes. One such co-operative, the Yaneshi Forestry Co-operative in the Peruvian Amazon, illustrates what is possible.[45]

The Yaneshi co-operative was established by a number of Peruvian Indians to manage their forest in the best way possible. Many land uses are inhibited by poor soil conditions, just like in most other rainforests. Their main activity is the shifting cultivation of manioc, rice and maize and the produce is processed locally. The main emphasis is on forest regeneration to ensure long-term sustainability and survival, and to generate enough income. Net annual returns of £3,500 per hectare harvested and processed have been achieved.

The Peruvian co-operative is a model which could be followed in many other rainforest areas. It provides employment, produces sustained yields from the forest and protects the cultural integrity of the local people.

6.7 CONTROLLING THE TROPICAL TIMBER TRADE

We saw in Chapter 3 that wood from tropical forests has many uses, including fuel for cooking and heating (see section 3.2). But by far the greatest pressure is for tropical hardwoods to meet consumer demand in developed countries (see section 3.6), which gives rise to what has been called 'the tropical chainsaw massacre'.[46] Very little of the hardwood traded on the international markets is derived from sustainable sources (by selective logging). An estimated 5 per cent of Britain's imports is sustainably produced.[47]

6.7a Timber trade

Most hardwood log exports are from South-East Asia to Japan (Figure 6.5a). South-East Asia also produces most of the tropical sawn log exports (Figure 6.5b) and the tropical plywood and veneers (Figure 6.5c). Japan has switched its main suppliers of hardwood as some areas have started to get logged out and governments have started to introduce bans on the export of logs. During the 1970s, for example, most came from Indonesia (Figure 6.4) but total imports fell during the 1980s when most came from Sarawak and Sabah.[48]

Clearly one approach to reducing clearance of the rainforests is to reduce the amount of hardwood exported from the tropical countries. Various options are available to do this, including imposing heavy taxes

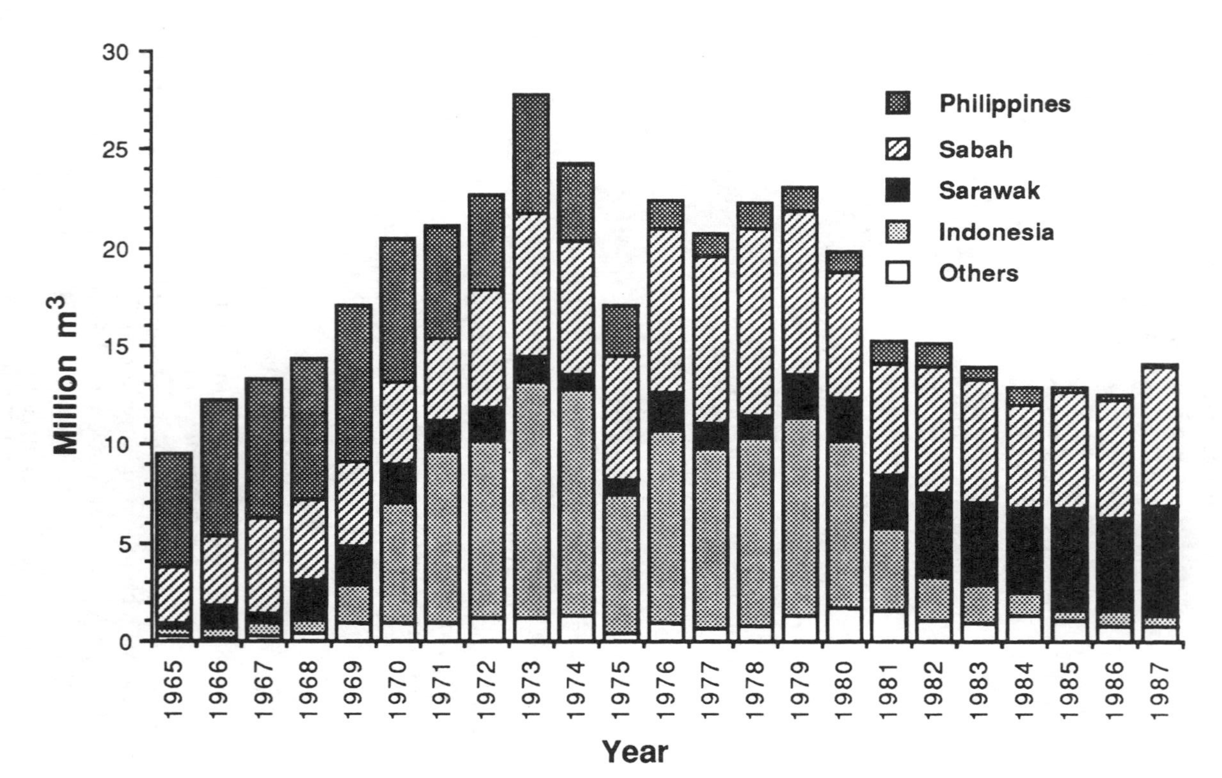

Figure 6.4 Main global flows of tropical timber, showing (a) hardwood logs, (b) sawn hardwood and (c) plywood and veneer.

Source: after Lewis (1990b)

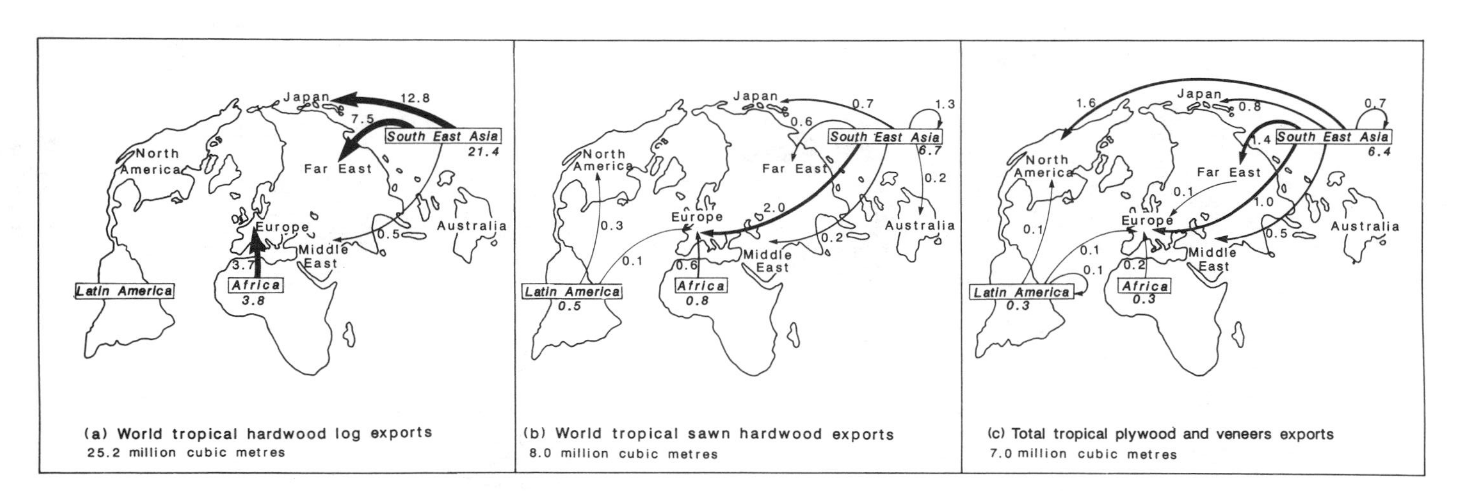

Figure 6.5 Changes in the supply of tropical hardwood to Japan between 1965 and 1987.

Source: after Lewis (1990b)

on imported tropical forest products and outlawing sales of tropical hardwoods from non-sustainable sources.

Voluntary schemes are preferable, and some countries have stopped importing wood from the rainforests. After the WWF and IUCN launched their Tropical Forests Campaign in 1982, for example, Switzerland abandoned the import of tropical hardwoods for coffins in favour of softwoods or temperate hardwoods.[49]

Other countries have stopped the exploitation and export of tropical wood. This mostly happens to protect surviving rainforests, as was the case when the government of Laos announed a total ban on all logging and timber exploitation in September 1991. Sometimes bans are imposed to decrease environmental disruption caused by deforestation (see section 4.4). In January 1989, for example, Thailand banned all logging in the wake of serious floods in southern Thailand that killed 350 people in November 1988 and were blamed mainly on illegal logging.

Some countries have pledged to stop exporting wood but have not in fact done so. President Sarney announced in mid-1988 that the export of raw timber from Brazil was to be banned, for example, yet by the end of the year it emerged that nearly 40,000 m^3 of tropical timber logs had been exported to Japan and up to 500,000 m^3 was expected to be exported in 1989.

Timber exports are sometimes too important a component of the national economy for a country to impose a blanket ban without facing serious economic consequences, and in such cases tougher controls might be applied to such exports without stopping them altogether. Faced with dwindling timber reserves, the Malaysian federal government tightened up on log exports in April 1991 and took control of log exports from the state of Sabah.

Two of the world's main tropical timber exporting countries – Indonesia and Malaysia – have repeatedly attacked calls from developed countries for a ban on imports of tropical hardwoods. They argue that it is unfair for western environmentalists to blame individual Asian countries for what are really global problems, and they insist that both countries have taken steps to sustain forest resources.

Rather than a total ban on the import and export of *all* tropical timber, some critics argue, efforts should be made to encourage or enforce trade in sustainable timber supplies. In May 1989, for example, the European Parliament passed a series of measures designed to regulate the European Community's US$5 billion a year imports of tropical woods and timber products and to create a Tropical Timber Management Fund from the EC's US$1.2 billion aid budget. It was agreed to give quotas to countries which export tropical timber and other forest products to the EC in proportion to the amount of timber each can produce sustainably; a country would receive compensation if it reduced its export volume. Financial aid would be available to support national forest management plans. The EC also agreed to stop all imports from countries which refuse to comply with the sustainable management obligations, after an agreed number of years.

6.7b Pressure groups and consumers

Consumer power is a highly potent force in rainforest clearance and campaigns by international environmental groups concerned about the future of the forests have targeted western consumer interest. In 1987, for example, Friends of the Earth in Britain launched a tropical hardwood consumer campaign.[50] Retailers, manufacturers, importers, architects, local authorities and other users of tropical hardwoods were lobbied to adopt a code of conduct ensuring that Britain trades only in well-managed, ecologically sound timber products.

The code of conduct, agreed in principle by the UK Timber Trade Federation, seeks to ensure that consumer countries will only import timber from sustainably managed areas.[51] It also proposes that the timber trade should create a tropical forest conservation and management fund, supported by an import surcharge on tropical timber.

By 1989 a hundred British companies (including national retail chains such as Habitat, Laura Ashley and the Body Shop) had agreed to adopt the code and stop using tropical hardwoods unless they came from properly managed forests.[52] More than thirty local authorities around Britain had committed themselves to adopt policies against the use of non-sustainably produced tropical hardwoods.

One strategy adopted by some timber retailers to capitalise on consumer interest in sustainable supplies of woods has been to use so-called 'eco-labels' which claim that the wood comes from sustainable logging operations. Many such claims are untrue and misleading according to a June 1991 WWF study, which calls upon retailers of tropical timber to check the facts and be honest in their statements. Even where labelling is used properly it creates conflict. In 1989 the International Tropical Timber Organisation (see section 6.8) proposed the introduction of labelling to distinguish timber from sustainable and non-sustainable sources; but Malaysia vetoed the idea, dismissing it as a trade barrier to protect the industrialised countries' own timber trade.

Lobbying by pressure groups can clearly have a significant impact on consumer awareness and behaviour, and many environmentalists believe that it is at this level that the most urgent changes are needed. But governments and many international agencies also recognise the need to control all aspects of the production and trade in tropical timber. This can only be done by international agreement and co-operation.

6.8 INTERNATIONAL TROPICAL TIMBER AGREEMENT (ITTA)

The International Tropical Timber Agreement (ITTA) is an economic treaty designed to get producers and consumers of tropical timber to co-operate to support the growth of the industry through improved forest management, timber processing and marketing. It also has a clear conservation objective, which makes it unique among international trade agreements.

The agreement has been signed and adopted by 42 countries (18 producers and 24 consumers) which account for 70 per cent of all tropical forests and 95 per cent of tropical timber exports.[53] It is supported by WWF, IUCN and other international conservation organisations.

ITTA is run by the International Tropical Timber Organisation (ITTO) under the aegis of the United Nations Convention on Trade and Development (UNCTAD). Headquarters for the organisation were established in Yokohama, Japan (the world's leading importer of tropical timber), in 1986.

6.8a Activities

ITTA and ITTO are designed to offer practical and financial support to exporter countries, with twin aims of conservation and sustainable management of the tropical forests. Funding is being made available for protected forest areas, to cover the costs of appointing forest ecologists, and finance reforestation programmes.

Some ITTO projects are looking at ways of developing and modifying logging systems to reduce their environmental impact. In 1989, for example, an ITTO team carried out a detailed study of logging activities in the tropical rainforests of Sarawak and recommended various steps to ensure that the forests there were better managed and conserved. A 1991 ITTO report recommended a reduction in timber production in Sarawak from 10 million m^3 a year to 9 million.

Other projects seek to develop better techniques for reforestation and forest management, particularly the phasing out of primary forest logging in favour of sustainable management. For example, the first meeting of ITTO (in December 1987) agreed to provide an initial $100,000 for a pilot project to encourage the commercial expansion of cut and sawn timber on a sustainable basis in western Amazonia, together with plywood production and small crafts.

ITTO also takes seriously its responsibilities for organising the tropical timber trade and for helping to fix better prices for timber.

6.8b Evaluation

ITTO has had a mixed reception. As early as 1988 WWF (and others) argued that it was failing to address the real problem which is the gross under-valuation of tropical timber on world markets (mainly because environmental costs are not reflected in timber prices).

ITTO has also been criticised for spending too much time monitoring the flow of tropical timber around the world (Figure 6.4), and critics argue that its main aim is to protect the timber trade rather than protect forests and their peoples. This is certainly true, but forest conservation *is* a major objective of ITTA and there are encouraging signs from the early years of ITTO's operations.

Certainly the organisation has played a significant role in conservation by encouraging research on improved reforestation and forest management. One ITTO study is investigating ways to rehabilitate 35,000 km^2 of tropical rainforest in the Indonesian province of East Kalimantan, for example.

ITTO has also created a framework in which producer and consumer countries can work together towards the sustainable management of rainforests. In 1990 the seventh session of ITTO agreed a plan of action which set the year 2000 as a target date by which all tropical forest timber should be harvested sustainably, and it drew up a set of guidelines on achieving sustainability, to be implemented in forestry policies of all tropical timber producing countries.

Within its commitment to sustainable management, the ITTO framework recognises that logging in some areas must be allowed to continue, at least for the foreseeable future. Thus, for example, in November 1990 the organisation approved the continued logging of key tropical forests in Malaysia and Amazonia.

One particular success has been in encouraging producer countries to maintain or increase timber production without clearing more natural forest areas wholesale. In some cases – in Cameroon, Indonesia, Colombia and the Philippines, for example – previously non-commercial tree species are now being used for pulpwood.[54] Other countries have established large and successful plantations of fast-growing tree species. Zambia, for example, expects to meet all its timber needs from such crops by the year 2000.

6.8c Bans on commercial logging

A number of exporter countries have introduced bans on commercial logging or on the export of tropical timber (see section 6.7). Not all of these bans have been a direct result of ITTA; Indonesia, for example, banned the export of whole logs in 1985. But there is no doubt that ITTO has raised awareness of the problems of wholesale clearing and of the lack of sustainability in conventional logging practices, and encouraged many exporter countries to re-evaluate the way they use their rainforest capital assets.

In 1988 Ghana and the Ivory Coast banned the logging of fourteen types of tropical hardwoods, and Thailand banned commercial logging in 1989.[55]

The Philippines government has approved a bill which bans tree cutting in 64 of its 73 provinces, although it is proving difficult to enforce on the ground. About twice as many logs are illegally exported to Japan each year as are officially allowed.

6.9 TROPICAL FORESTRY ACTION PLAN (TFAP)

The early 1980s witnessed a growing awareness of the pace and causes of tropical deforestation (see Chapters 2 and 3) and the emergence of a better understanding of its consequences for ecology, environmental systems and world climate (see Chapter 4). While some tropical governments were starting to take action to conserve their own forests, deforestation overall was getting faster and more damaging.

6.9a Evolution of the plan

International conservation agencies were among the best placed groups to try to halt the rainforest clearance, and in 1982 the World Wildlife Fund (WWF, later renamed the World Wide Fund for Nature) and the International Union for the Conservation of Nature (IUCN) launched a Tropical Forest Campaign.[56] The campaign was designed to increase international understanding of deforestation and provide a vehicle for bringing together government leaders from the countries involved (producers and consumers).

In 1985 an International Task Force was set up under the umbrella of the Tropical Forest Campaign. It was supported by the World Bank, the World Resources Institute (WRI), the UN Development Programme (UNDP) and the UN Food and Agriculture Organisation (FAO), as well as WWF and IUCN.

The Task Force drafted a global action plan in 1985, called 'Tropical Forests: A Call for Action'. The programme would involve fifty-six countries, last for five years and cost an estimated US$8 billion.[57] It set itself a series of ambitious global objectives, including preventing further deforestation, saving remaining tropical forests and averting an imminent fuelwood crisis.

Those who drafted the plan realised from the outset that the commitment of all countries concerned was a prerequisite for any effective action, so major efforts were made to unite them under an agreed set of objectives and priorities.

A key step in the process of working out a viable plan which would receive wide endorsement was a meeting held at Bellagio, Italy, in July 1987. Top representatives from governments, development agencies, industry and non-governmental agencies met there to discuss commitment and support for the plan, which by then had received $1 billion in committed support. There were two outcomes from Bellagio – an agreed Tropical Forestry Action Plan (TFAP) and a broad enough base of support to suggest that it could be made to work.

6.9b The plan

The aim of the plan was to 'address deforestation on a broad front', and it sought to do this in five complementary ways:

(a) conservation of tropical forest ecosystems would ensure the survival of remaining rainforests
(b) agro-forestry would be promoted because it would increase the sustainable use of existing forests
(c) reforestation would replace at least some of the cleared forests and help to offset some of the worst environmental and climatic consequences of clearance
(d) supplies of fuelwood would be increased by replanting schemes and better forest management
(e) the institutional framework for forestry research, training and extension work within the developing countries would be strengthened to make it more suitable for managing the forest resources of the tropics.

The plan was multi-dimensional and sought to conserve, restore and encourage sustainable use of rainforests. This was the first concerted effort to tackle the deforestation problem on such a broad front and involving so many countries.

The architects of the plan envisaged four main steps to bring about the five key objectives.[58] First, all of the tropical nations with rainforests must put a price tag on deforestation and include it in their economic and development plans. Second, national forestry reviews must be carried out in all tropical countries to assist in long-term planning. Third, incentives should be given to encourage private investment in reforestation schemes. Fourth, the involvement of non-governmental organisations (NGOs) and community groups in the scheme should be encouraged.

6.9c Implementation

The plan was ambitious but the participating countries adopted its recommendations and set about putting them into practice. By 1990

National Forestry Action Plans were being prepared in sixty countries, with technical and financial help from international development agencies. These would help the governments concerned to assess the status of their forests, estimate industrial potential and evaluate training needs.

The plan had several stages to be implemented over a number of decades. The most immediate benefits would come from 'debt-for-nature' swaps, designed to enable tropical countries to relieve some of their international debt burdens by pledging to conserve rainforests (see section 6.10). In the longer term, benefits would also accrue from reforestation schemes, the phasing out of destructive development schemes and reformation of development strategies in tropical countries.

6.9d Development schemes and strategies

Whilst reforestation can restore forest cover in cleared areas, its potential is limited for various reasons, particularly in the moist tropics. Moreover, debt-for-nature swaps (see section 6.10) would be difficult if governments were continually put under pressure to use forest areas for development projects.

Slowing down or stopping the forces which fuel deforestation is a preferable objective, because prevention is always better than cure. TFAP recognises the need to phase out development schemes which involve forest clearance (such as logging and dam construction) in order to preserve as much virgin forest as possible. This approach also has the advantage that forest peoples no longer need to be displaced from their natural habitat.

Phasing out damaging development schemes requires economic sacrifices from the very countries which are generally least able to make them. Consequently a reformation of development strategies in these countries is required, which still allows them to prosper without selling off their biological capital (the forests).

Deforestation often results from development strategies which encourage developing countries to buy manufactured consumer goods and technical devices from the developed world. Means must be found of enabling the developing countries to import goods they can pay for without squandering more of their irreplaceable rainforest estate.

The plan also recognises the need for international banks and aid agencies to switch their lending away from projects which lead to forest clearance and towards non-damaging, sustainable projects.

The task of reforming the basis of development in many tropical countries is immense and it is urgent. This is one of the most important solutions to the problems of deforestation, which requires great political goodwill and trust, along with public understanding and support. It is not an easy path to embark on.

6.9e Evaluation

The Tropical Forestry Action Plan has had a mixed reception. One of its major successes has been to encourage the involvement of non-government organisations (NGOs) in decision-making about the future of the forests, because they provide a grassroots perspective on local needs and opportunities.

At the other end of the scale, it has also successfully brought together development and aid agencies like FAO, the World Bank and the UN Development Programme. Co-ordinating the assistance such agencies provide to forestry should help to prevent wastage and duplication of funding, making better use of the available resources.

TFAP has also been effective in encouraging developed countries to provide financial and material support for forestry initiatives in developing tropical countries. In 1986, for example, France, West Germany and the Netherlands agreed to double their aid to forestry.[59]

Raising public and government awareness of the tropical deforestation problem, which are very difficult to quantify, has doubtless been a major achievement of the plan.

But it is not all good news. The plan has been heavily criticised by environmental groups round the world as being underfunded and badly directed.[60]

In March 1990 a coalition of interested parties (the World Rainforest Movement, the *Ecologist* magazine and Friends of the Earth) concluded that the plan is totally flawed and urged that international financing of it should stop pending a radical review. They argued that the plan fails to address the social and political roots of deforestation, and pointed out that the National Forestry Action Plans are dominated by the concerns of conventional forestry organisations. Several months later the World Resources Institute[61] also called for a major reform of the plan as well as suggesting a Global Forest Convention to balance exploitation with conservation.

One problem is the top-down approach embodied in the plan, which reinforces the view of some critics that it reflects imperialism rather than conservation. This is borne out in the allocation of funds during 1985, when only 8 per cent of the TFAP budget was allocated to the conservation of ecosystems while much more was spent on industrial use, agro-forestry, bureaucracy and other land uses.[62]

One serious flaw in the plan is that it did not involve or consult with forest dwellers. Consequently it fails to recognise let alone meet the needs and aspirations of forest people. The UN Food and Agriculture Organisation has since issued guidelines for the national forestry reviews which stress the need to consider the roles of indigenous people, women and NGOs.[63]

Critics argue that there is an unwritten but underlying presumption to the plan, which is that the prime causes of deforestation are poverty, over-population and ignorance. It blames landless peasants and firewood collectors for deforestation, rather than commercial operators. Blaming the poor, the critics point out, has given an excuse for current damaging development policies to continue.

Another weakness is that the plan did not put forward measures to prevent projects which cause damage, thereby ensuring that deforestation continues. It has even been argued that the plan trivialises the effects on the forests and their peoples of large-scale development.[64]

The plan has also been attacked as misguided, because none of the forty-two tropical countries which were formulating Forestry Action Plans in 1988 were seriously considering forest restoration schemes.[65]

6.9f Reformation

In October 1990 the UN Food and Agriculture Organisation (FAO) agreed to reform the plan in the face of the mounting international criticism and the obvious failure of the original formulation. By late summer 1990, for example, only 6 out of more than 75 interested countries had forestry plans inaugurated under TFAP.

The reform was far-reaching. FAO agreed to strengthen the co-ordinating body and increase the emphasis on rapid response to requests for assistance. In March 1991 TFAP was renamed the Tropical Forestry Action Programme (conveniently retaining its established acronym), and its objective was revised to 'conservation and sustainable development of forestry resources in the interests of the country concerned and the global community'. The re-orientation of TFAP gave greater prominence to the major causes of deforestation and allowed the inclusion of forest-dwellers and non-government organisations (NGOs). The new programme will focus less on individual projects and more on broad policy changes, and is likely to be administered nationally rather than centrally by a small unit based in Rome.

6.10 DEBT-FOR-NATURE SWAPS

Many developing tropical countries owe large debts to developed countries, commercial banks and institutions such as the World Bank, having borrowed heavily to finance major development schemes. Most cannot service these debts or even manage the interest payments, and so they sacrifice their endowment of natural resources such as rainforests and use them as capital.[66]

The Tropical Forestry Action Plan and others have proposed that debt-ridden tropical countries realise the value of their ecological capital assets and literally trade areas of forest for their international debt.[67] 'Debt-for-nature' swaps involve the country's creditors selling the debt at a discount price to a conservation group or foreign government. The purchaser then waives the debt in exchange for an undertaking from the vendor country that it will set aside and manage an agreed area of rainforest.

The net effect is to reduce the national debts of countries that take part in conservation programmes, while at the same time enabling them to slow down or stop deforestation.

At the end of 1987 WWF in the United States promoted two debt-for-nature swap initiatives, in Ecuador and Costa Rica. In Ecuador the plan was to buy US$1 million in Ecuadorian debt, which would yield up to US$6 million in conservation benefit. The money was destined to strengthen the country's system of National Parks and reserves. The Costa Rica scheme involved raising money to buy a remnant of rainforest as a seed source to replant over 700 km^2 of degraded pastureland.

Some countries have adopted such debt-swapping strategies with positive results for rainforest conservation. Conservation International, a United States organisation, negotiated to buy US$650,000 worth of Bolivia's debt at a discounted rate of US$100,000. The debt was then written off in exchange for the Bolivian government undertaking to set aside 15,000 km^2 of Amazon rainforest.[68]

Some national debts are simply too large for debt swaps alone to be a realistic solution. Brazil, for example, owed US$115 billion to western banks by the close of the 1980s. But as part of a broader debt- and development-restructuring package, such schemes show promise. Brazil radically altered its policy towards the rainforest at the start of the 1990s (away from development and towards conservation) and saw the use of conservation to bargain for reduction in foreign debt as a way of solving two of its most intractable problems at the same time.

In February 1991 the Mexican government agreed a debt-for-nature swap with Conservation International, who agreed to purchase and write off US$4 million worth of Mexican debt from foreign creditors (at the discounted rate on the secondary market of about $1.8 million). In return the Mexican government agreed to invest US$2.6 million in rainforest conservation, with a particular focus on the Selva Lacandoni (the largest remaining tropical rainforest in North America).

Not all debt-for-nature swaps are welcomed by all parties. The Bolivian government implemented a debt swap in 1987 which led to the designation of half of the country's Chimanes forest as a 'permanent production zone' which would be available for sustainable forestry; the rest of the area was to be used as a research area. But in mid-1990 tribal Indians in that area challenged the very premise of the designation, arguing that logging companies have continued extracting timber from the area, driving away wildlife and polluting rivers as well as felling the forest.

Critics of the debt-for-nature swaps dismiss them as a modern form of imperialism. They also argue that the swaps will have little impact on deforestation overall, because setting aside small areas of forest may lead to greater exploitation of the unprotected areas. Whilst supporters concede this difficulty, they do point out that the swaps play a significant role as holding actions to gain time before more suitable policies can be implemented.

6.11 AID POLICIES AND INVESTMENT DECISIONS

Most observers point out that the best way to solve the problem of deforestation is to attack the root cause of the pressures on rainforests. Initiatives like those which tackle the international trade in timber or debt-for-nature swaps, for example, are genuine steps forward but they will not by themselves solve the problem.

6.11a Development and quality of life

There must be complementary attacks on the root problems of poverty, under-employment, food and energy deficiencies and uncontrolled population growth in developing tropical countries. This requires significant improvement in the economies of developing countries, so that they have a well-balanced economy that does not require wholesale development of the rainforest as a resource for short-term gain.

Tropical nations often favour destructive, large-scale development projects in their pursuit of development and they are encouraged and often funded by international agencies such as the World Bank. Many see the large-scale development projects as essential to produce enough money in the short term to be able to afford more sustainable development in the long term.

Deforestation, and attempts to stop it, must be viewed against a background of debt in the Third World. Norman Myers[69] has claimed that the pressure of international debt has served to promote cattle ranching in the Amazon Basin, has undermined logging restrictions in Ecuador, the Ivory Coast and Indonesia and has expanded the growing of cash crops on farmlands, thus pushing subsistence farmers into the more fragile environments in the Sahel.

Recent developments in Indonesia show how damaging development schemes, funded by international finance, are often accepted by developing countries faced with the desire if not the need for rapid economic development. Early in 1991, fuelled by its desire to become the world's top paper-producing nation, Indonesia embarked on a major expansion of its pulp and paper industry, involving the clear-felling of rainforests and planting of fast-growing species. Paper exports only began in 1987, yet by mid-1991 there were 41 paper mills. Over the next fifteen years the government plans to build another 56 large pulping mills, which could only be fed by a massive increase in pulpwood plantations. The plan is to clear-fell huge areas (up to 8,000 km^2 eventually) of lowland tropical rainforest in the south east of Irian Jaya (West Papua).

6.11b Local peoples

There is a growing consensus that local peoples should be encorporated into the management and development of the remaining rainforests. This would help to remove much of the responsibility for the fate of the forests from the hands of a small rich elite, and place it back in the hands of those to whom it rightly belongs. It would also answer the common criticism that control of the forests is often given to outside agencies whose main interest is the rate of return to be gained from cropping.

Most of the international investment inevitably benefits the investor more than the developing country which receives it. Japanese corporations have been criticised for financing the clearance of large areas of rainforest in Chile, which in 1989 alone provided around 2 million tonnes of wood chips for the Japanese market. Few if any of the benefits from this sort of scheme are enjoyed by local people.

6.11c Quantity versus quality of loan

Critics of international development aid stress that one of the main problems has been the overriding emphasis within the banks on *quantity* not *quality* of lending. For example, the World Bank committed US$443 million to the Polonoroeste scheme in western Brazil (see section 3.9) to fund road construction and farming by settlers, both of which required massive forest clearance. But the whole scheme was soon seen as a waste of environmental resources as well as valuable capital, because soils were simply not suitable for permanent cultivation.

A 1990 European Parliament report shows that European Community aid to underdeveloped countries is being spent on projects which directly destroy rainforests in Asia, Africa and South America. Community aid has helped to fund a road cutting through one of Zaïre's rainforests (which will increase production at a meat processing plant), and a timber extraction project in Equatorial Guinea (which will bring a fourfold increase in timber production in four years). EC funding worth US$257 million also went to the Carajas iron ore project in the Brazilian state of Para (see section 3.9).

6.11d Revision of objectives

Towards the close of the 1980s the multilateral development banks (such as the World Bank) and bilateral aid agencies (such as Britain's Overseas Development Administration) had learned the lessons of funding projects that destroy the rainforests, and they were starting to change their priorities and objectives.[70] The United States aid agency USAID has been instructed by the Congress to consider the effects on species diversity of the dams, roads and other developments it funds.[71] The World Bank and Inter-American Development Bank revised the basis on which they provide funding for projects which damage tropical rainforests. In November 1988, for example, the World Bank announced that it would not fund either Babaquara or Kararao (the two main Xingu River hydroelectric dams in Brazil) on environmental grounds.

From being the 'bad guy' in tropical deforestation the World Bank is starting to re-emerge as the 'good guy'. In July 1991 the Bank announced that it will no longer fund logging projects in tropical rainforests, and it launched a new policy which recognises the rights of forest dwellers and local people to continue with traditional ways of life. The Bank's new objectives include assisting reafforestation projects, bringing degraded land back into production, developing secondary forests and financing the policing of ancient forests – rather than funding damaging development schemes (see sections 3.9 and 3.10).

6.12 CHANGING FORTUNES IN AMAZONIA

Amazonia has long been a focal point of international concern about the survival of the rainforest, and recent events in Brazil illustrate many of the inherent complexities of arriving at realistic and effective solutions to the problems of forest clearance. In final analysis both political goodwill and economic viability are required. In practice, both change through time.

After many years of apparent indifference to international calls for a halt to Amazon clearance, towards the close of the 1980s a new awareness of the need to preserve the country's irreplaceable ecological heritage seemed to be emerging.

Late in 1988 President Sarney announced a suspension of the government tax breaks and subsidies that have made major Amazon development projects (such as cattle ranching; see section 3.8) economically viable. Other measures included a total ban on the export of logs, and the imposition of strict environmental controls over all future agricultural and industrial projects. A national educational programme was promised which would increase public understanding of the problems of deforestation. Structural changes were proposed, too, which would merge the forestry, fishery and environment agencies into one new department and create a new ecology and human rights department. Critics of the plan argued that it failed to address the key issues, such as land speculation, land tenure, population migration and highway construction.

By early 1989 some of the initial euphoria was starting to evaporate as some of the constraints became more apparent. One was opposition to external pressure, and it was manifest when a nationalist groundswell in Brazil started to voice strong objections to what was seen as international intervention in the affairs of a sovereign state. But there were powerful internal constraints, too, and it was clear (particularly from the uncontrollable gold rush by illegal miners; see section 5.5) that conditions in the interior of Amazonia were verging on anarchy. The Ministry of Justice in Brazil reported a serious lack of protection for the forest itself, as well as many violations of human rights, corruption, disrespect for the constitution and omission by government departments. All the evidence suggested that the government agency FUNAI lacked both the funds and the means to protect the Indians and the forest.

President Sarney reacted to the criticism by announcing a new environmental protection programme in April 1989. The 'Our Nature' scheme reflected national sovereignty and responsibility. Centrepiece of the scheme was a US$100 million mapping and zoning study designed to determine which parts of the Amazon Basin are suitable for development, what kind of development would best suit each region and what areas should be left alone. A series of new parks and reserves were envisaged, along with a better-equipped forest service and an educational campaign.

Amazonia is not confined to Brazil, and the other rainforest countries were well aware of the need to co-ordinate their policies and to co-operate in protecting the surviving forest. Leaders of all eight Amazon countries met together in May 1989 to discuss the future of the rainforest, and all eight Presidents signed a *Declaration of Amazonia* which called on the countries of the developed world to provide money for the preservation of the rainforest and for the economic development of the region (faced with a total debt of around US$400 billion which could not be repaid under existing conditions).

The need for international aid outweighed feelings of nationalistic pride, and in July 1989 Brazil accepted help (funding and expertise) from the UK to establish a major rainforest research programme. The programme had a number of practical ingredients, including the setting up of a biological reserve at Caxiuna near Belem for research into aromatic plants. Resources were also made available to help train Brazilian officials in how to explore the environmental impacts of projects. Other projects were planned on sustainable management of the rainforest, including reforestation, agro-forestry, forest regeneration and seed research and production. A further US$50 million in grants and low-interest loans was received from West Germany to build up the Brazilian Institute of the Environment and Natural Renewable Resources (IBAMA) and establish and maintain forest reserves.

Further progress came in March 1990 when President Sarney announced the creation of three extractive reserves covering a total of 16,000 km^2 in Amazonia for the use of rubber tappers. The Chico Mendes Extractive Reserve in Acre (9,705 km^2) will benefit around 7,500 tappers, the Rio Cajari Extractive Reserve in Amapa (4,816 km^2) will benefit around 5,000, and an unspecified number will benefit from the Rio Ouro Preto Extractive Reserve in Rondonia (2,045 km^2). Land within these reserves will continue to belong to the federal government, but long-term contracts are granted to local associations of rubber tappers to ensure sustainable use of the forest resources (it is in the long-term interest of the rubber tappers themselves to use their resources sustainably).

Remote sensing surveillance by the Brazilian National Space Research Institute (INPE) has yielded both bad and good news about the pace of Amazon clearance. Figures released by INPE in June 1990 showed that destruction of the Amazon rainforest has been much greater (8 per cent) than the Brazilian government had previously reported (5.6 per cent was announced in April 1989). But a March 1991 INPE report showed a recent decrease in the rate of forest burning, down by about 27 per cent between 1989 and 1990. The decrease was attributed to the effectiveness of the government's tax changes and the reduction in government subsidies for farming in Amazonia. Even the March 1991 report had mixed news, however, because INPE stressed that burning is still on the increase in some areas (particularly in Rondonia).

A change in attitudes towards Amazon clearance came the very next month, in April 1991, when the new President (Collor) announced a significant U-turn in government policy towards Amazonia. He reintroduced the financial incentives for deforestation which his predecessor (Sarney) had abolished in 1990. He also granted funding for the BR364 highway which would cut through the western Amazon – a highway scheme which the World Bank had previously refused to fund on environmental grounds.

Such was the international outcry at this reversal of fortune for the Brazilian rainforest, and such was the economic pressure on President Collor to revise his bullish thinking, that he did so with two months. In June 1991 the President issued a decree stating that government subsidies for agricultural development (including cattle ranching) would be withheld from any projects which involve deforestation. The government also agreed that an annual $100 million of its foreign debt could be converted into funds for environmental protection under debt for nature swaps. The head of FUNAI, the controversial government agency charged with the protection of the interests of native Indian people, was sacked in the same month, allegedly for failing to demarcate the lands of the Yanomami Indians at risk from gold miners and other threats.

The struggle for the survival of the Amazon continues unabated, and the eyes of the world remain on Brazil and its neighbours. Through time the nature and severity of the pressures on Amazonia will alter, and whilst one battle might have been won in the early 1990s (particularly in removing *some* of the economic incentives to clear rainforest), very clearly the war is far from over.

6.13 CONCLUSIONS

Having examined many of the suggested solutions to the problem of tropical deforestation, it is difficult to avoid the conclusion that the problem is immensely complex and there is no simple or single solution.

A wide variety of solutions have been suggested,[72] including more research and development on tropical forest resources, more and better education about the forests, more and better forest conservation and restoration schemes, more widespread use of debt-for-nature swaps, increased commodity prices for and import restrictions on timber, and increased co-operation between countries to seek viable and sustainable uses of rainforests. Without doubt all of these measures – and more – are required, and without delay.

In many ways the problems of the rainforest are microcosms of the problems of the world, because they reflect the tensions and interplay between the powerful and the powerless, the rich and the poor, the north and the south. As a result some critics of existing attitudes and policies towards the forests have joined the call for a new world economic order, which would be more sustainable, less environmentally damaging and more equitable. Prince Charles, in his Rainforest Lecture at Kew Gardens,[73] emphasised the need to change our attitude to the planet in ways which would lead to changing to forms of development which are sympathetic to the rainforests.

The time-bomb of ecological, environmental, climatic and human damage caused by deforestation continues to tick, and the problem of tropical rainforest clearance must remain a priority within international politics.

NOTES

1 THE TROPICAL RAINFOREST: HISTORY AND ENVIRONMENT

1 Darwin (1845, pp.22–4).
2 Other useful sources include Richards (1970, 1976), Caufield (1985), Flenley (1979), Sutton *et al.* (1983), Tucker and Richards (1976), Silcock (1989) and Whitmore (1990).
3 World Resources Institute (1988).
4 World Resources Institute (1988).
5 Independent Commission on International Humanitarian Issues (1986).
6 Sternberg (1988), Colinvaux (1989).
7 Grainger (1980, 1984).
8 Hammond (1977).
9 World Resources Institute (1988).
10 World Resources Institute (1988).
11 World Resources Institute (1988).
12 World Wide Fund for Nature (1988).
13 Moore (1986).
14 Myers (1988a).
15 Moore (1986).
16 Flenley (1979), Elsworth (1990).
17 Walker (1986).
18 Sterling (1973).
19 Botkin and Keller (1987, p.160).
20 Elsworth (1990).
21 Eyre (1981, p.26).
22 Goudie (1984).
23 Richards (1973, 1976), Sterling (1973), Goudie (1984), Collinson (1977), Eyre (1968, 1971).
24 Walker (1986).
25 Moore (1986), Erwin (1988).
26 Ramade (1984).
27 Walker (1986).
28 Goudie (1984).
29 Richards (1973).
30 Friends of the Earth (1989c).
31 Jordan (1985).
32 Caufield (1985).

33 Friends of the Earth (1989c), Hecht and Cockburn (1990).
34 Smith (1980).
35 Young (1974), Nortcliff (1989).
36 Nortcliff (1989).
37 Caufield (1985).
38 Salati and Vose (1984).
39 Goudie (1984).
40 Holm-Nielson *et al.* (1989).
41 Richards (1973).
42 Friends of the Earth (1989a).
43 Sayer and Stuart (1988).
44 International Earthcare Centre (1983, p.71).
45 Friends of the Earth (1989a).
46 World Wide Fund for Nature (1988), Elsworth (1990).
47 Friends of the Earth (1989a).
48 Friends of the Earth (1989c).
49 Goldsmith and Hildyard (1990).
50 Friends of the Earth (1989c).
51 Sterling (1973).
52 Data from Sterling (1973), Friends of the Earth (1989a).
53 Myers (1985b).
54 Data from Sterling (1973), Friends of the Earth (1989a).
55 Goudie (1989).
56 World Wide Fund for Nature (1988).
57 Elsworth (1990).
58 Friends of the Earth (1989a).
59 Cross (1990).
60 World Wide Fund for Nature (1988).
61 Stott (1978), Sioli (1985), Colinvaux (1989), Whitmore (1990).
62 Colinvaux (1989).
63 Moore (1990), Park (1992).

2 DESTRUCTION OF THE RAINFOREST: RATES OF LOSS

1 Myers (1990).
2 Friends of the Earth (1989c).
3 Denevan (1973).
4 Flenley (1979).
5 Food and Agriculture Organisation (1982).
6 Earthlife Foundation (1985).
7 Myers (1980a), Salati and Vose (1983), Grainger (1984), Furtado and Ruddle (1986).
8 Fearnside (1982), Caufield (1985).
9 Myers (1989a, 1989b), Friends of the Earth (1989c).
10 World Wide Fund for Nature (1988).
11 Friends of the Earth (1989c).
12 Friends of the Earth (1989a).
13 Tyler (1990a).
14 Sting and Dutilleux (1989).
15 National Academy of Sciences (1982).

16 World Resources Institute (1988).
17 World Resources Institute (1990).
18 Tucker *et al.* (1984), Malingreau *et al.* (1985).
19 Myers (1983).
20 United Nations Environment Programme (1987).
21 World Resources Institute (1990).
22 Bunker (1980), Drozdiak (1982), Lemonick (1987), Ellis (1988), Linden (1989).
23 Scott (1986, 1989), Hurst (1989).
24 Myers (1988a).
25 Friends of the Earth (1989c).
26 Elsworth (1990).
27 United Nations Environment Programme (1987).
28 Walker (1986), Elsworth (1990) .
29 World Wide Fund for Nature (1988).
30 World Wide Fund for Nature (1988, p.2).
31 *The Economist* (1988)

3 CAUSES AND PROCESSES OF CLEARANCE

1 Allen and Barnes (1985).
2 Eckholm, Foley and Bernard (1984), Foley (1985).
3 World Wide Fund for Nature (1988).
4 World Bank (1984).
5 World Resources Institute (1986).
6 Ooi Jin Bee (1983, p.47).
7 Elsworth (1990).
8 Bromley (1972), Hiraoka and Yamamoto (1980).
9 Goldsmith and Hildyard (1990).
10 Goldsmith and Hildyard (1990).
11 International Earthcare Centre (1983, p.72).
12 Friends of the Earth (1989c).
13 Plumwood and Routley (1982).
14 Kellogg (1963).
15 Denevan (1966), Clarke (1976), Janzen (1973), Nortcliff (1989).
16 Melillo *et al* (1985).
17 McIntyre (1988).
18 Caufield (1985).
19 Independent Commission on International Humanitarian Issues (1986).
20 World Wide Fund for Nature (1988, p.2).
21 Secrett (1986).
22 Caufield (1984, p.40).
23 Survival International (1989).
24 Goldsmith and Hildyard (1990) .
25 Caufield (1985).
26 Goldsmith and Hildyard (1990).
27 Friends of the Earth (1989a).
28 Caufield (1985, pp.143–4).
29 Friends of the Earth (1989c).
30 *The Economist* (1988, p.26).
31 Friends of the Earth (1989c).
32 Wells (1989).

33 Fearnside (1983).
34 Vesiland (1987).
35 Friends of the Earth (1989c).
36 *The Economist* (1988).
37 Goldsmith and Hildyard (1990).
38 Tyler (1990a).
39 Tyler (1990b).
40 Caufield (1985).
41 Elsworth (1990).
42 World Resources Institute (1987).
43 Friends of the Earth (1989a).
44 Elsworth (1990).
45 World Wide Fund for Nature (1988).
46 Scott (1986).
47 Swinbanks (1989).
48 World Wide Fund for Nature (1988).
49 Caufield (1985).
50 World Wide Fund for Nature (1988).
51 World Wide Fund for Nature (1988).
52 World Resources Institute (1988).
53 Sting and Dutilleux (1989).
54 Ooi Jin Bee (1983).
55 Sting and Dutilleux (1989).
56 World Wide Fund for Nature (1988).
57 Plumwood and Routley (1982).
58 World Wide Fund for Nature (1988).
59 World Resources Institute (1987).
60 Caufield (1985).
61 Tyler (1990b).
62 World Wide Fund for Nature (1988).
63 Myers (1989b).
64 World Wide Fund for Nature (1988).
65 Myers (1989b).
66 World Wide Fund for Nature (1988, p.3).
67 *The Economist* (1990).
68 Plumwood and Routley (1982).
69 World Wide Fund for Nature (1988).
70 Myers (1988d).
71 Caufield (1985).
72 Goldsmith and Hildyard (1990).
73 Russell (1942), Linden (1989).
74 Drozdiak (1982).
75 *The Economist* (1988).
76 Friends of the Earth (1989a).
77 Button (1988).
78 Friends of the Earth (1989c).
79 Smith (1981).
80 Fearnside (1983).
81 Townsend (1990).
82 Fearnside (1983).
83 Townsend (1990).
84 Friends of the Earth (1989c).
85 Friends of the Earth (1989c).

86 Elsworth (1990).
87 Elsworth (1990).
88 Myers (1981, 1988a), Myers and Tucker (1967).
89 Goldsmith and Hildyard (1990).
90 Caufield (1985).
91 Hecht (1990).
92 Fearnside (1983).
93 Caufield (1985).
94 Goodland (1980), Fearnside (1987).
95 Goodland and Brookman (1977).
96 Smith (1981).
97 Plumwood and Routley (1982).
98 Smith (1981).
99 Friends of the Earth (1989c).
100 Friends of the Earth (1989c).
101 Friends of the Earth (1989c).
102 Lewis (1990a).
103 Caufield (1982, 1983).
104 Fearnside (1986), Hall (1989), Townsend (1990).
105 Townsend (1990).
106 Goldsmith and Hildyard (1990).
107 Friends of the Earth (1989c).
108 Caufield (1985).
109 Caufield (1985).
110 Tyler (1990b).
111 Fearnside (1989).
112 Fearnside (1989).
113 Friends of the Earth (1989c).
114 Friends of the Earth (1989c).
115 Fearnside (1987), Hecht and Cockburn (1990).
116 Goldsmith and Hildyard (1990).
117 Friends of the Earth (1989c).
118 Fearnside (1989).
119 Friends of the Earth (1989a).
120 Friends of the Earth (1989c).
121 Friends of the Earth (1989c)

4 IMPACTS AND COSTS OF DESTRUCTION

1 Gomez-Pompa *et al.* (1972).
2 Bowonder (1987).
3 Cross (1990).
4 Gomez-Pampa *et al.* (1972) .
5 Myers (1980b), Ehrlich and Ehrlich (1982).
6 Goldsmith and Hildyard (1990).
7 Lovejoy (1989).
8 Friends of the Earth (1989a).
9 World Wide Fund for Nature (1988).
10 Moore (1986).
11 Friends of the Earth (1989a).
12 Friends of the Earth (1989c).
13 World Wide Fund for Nature (1988).

14 Quoted by McIntyre (1988, p.55).
15 Myers (1988a).
16 World Wide Fund for Nature (1988, p.2).
17 Myers (1989a, p.17).
18 Myers (1988d).
19 Earthlife Foundation (1985).
20 Elsworth (1990).
21 World Wide Fund for Nature (1988).
22 Myers (1988d).
23 Friends of the Earth (1989a).
24 Elsworth (1990).
25 Friends of the Earth (1989a).
26 World Wide Fund for Nature (1988).
27 Friends of the Earth (1989a).
28 Friends of the Earth (1989c).
29 World Wide Fund for Nature (1988).
30 Prescott-Allen and Prescott-Allen (1988), Smith and Schultes (1990).
31 Friends of the Earth (1989a).
32 Myers (1988b).
33 World Wide Fund for Nature (1988, p.3).
34 Friends of the Earth (1989a).
35 *The Economist* (1988).
36 World Wide Fund for Nature (1988).
37 Myers (1989a).
38 Friends of the Earth (1989a).
39 Grainger (1980).
40 Friends of the Earth (1989a).
41 World Wide Fund for Nature (1988).
42 Elsworth (1990).
43 Friends of the Earth (1989a).
44 World Wide Fund for Nature (1988).
45 Myers (1988b, 1988d).
46 Smith (1981).
47 Schneider (1989).
48 Fox (1976).
49 *The Economist* (1989).
50 Myers (1988d).
51 Friends of the Earth (1989c).
52 Friends of the Earth (1989a).
53 Wells (1989).
54 World Wide Fund for Nature (1988).
55 Myers (1988d).
56 McDonagh (1986).
57 World Wide Fund for Nature (1988).
58 Earthlife Foundation (1985).
59 Gentry and Lopez-Parodi (1980).
60 Goldsmith and Hildyard (1990).
61 Dickenson (1982), Myers (1988c, 1989b).
62 Goudie (1984).
63 Henderson-Sellers and Gornitz (1984).
64 Schneider (1989).
65 World Wide Fund for Nature (1988).
66 Salati and Vose (1984, p.130).

67 Myers (1988c).
68 Myers (1988c, 1989b).
69 Myers (1988a).
70 Myers (1988a).
71 Walker (1986).
72 Friends of the Earth (1989c).
73 Salati and Vose (1983).
74 Friends of the Earth (1989a).
75 Bunyard (1985).
76 Goldsmith and Hildyard (1990).
77 Goudie (1984).
78 Bunyard (1985).
79 Sedjo (1989).
80 World Wide Fund for Nature (1988, p.2).
81 Bolin (1977), Woodwell *et al.* (1978), Revelle (1982), Jager (1986), Detwiler and Hall (1988).
82 Friends of the Earth (1989c).
83 Independent Commission on International Humanitarian Issues (1986).
84 Schneider (1989).
85 Friends of the Earth (1989c).
86 Myers (1989b).
87 Walker (1986).
88 Myers (1989b).
89 Cook *et al.* (1990).
90 Fearnside (1989).
91 Goudie (1984).
92 Bunyard (1985), Hekstra (1989), Myers (1989c), Tyler (1989), Park (1991).
93 World Wide Fund for Nature (1988, p.2)

5 FOREST PEOPLES

1 World Resources Institute (1990).
2 Sting and Dutilleux (1989).
3 Mills (1990).
4 Rapaport (1976), Sioli (1985).
5 Caufield (1985).
6 Caufield (1984, p.87).
7 Reichel-Dolmatoff (1977).
8 Caufield (1984).
9 Friends of the Earth (1989a).
10 World Wide Fund for Nature (1988).
11 World Wide Fund for Nature (1988).
12 Gray (1990).
13 Bunyard (1974).
14 Friends of the Earth (1989a).
15 Caufield (1985).
16 *The Economist* (1988).
17 Fearnside (1989).
18 Tyler (1990b).
19 Friends of the Earth (1989a).
20 McIntyre (1988).
21 Independent Commission on International Humanitarian Issues (1986).

22 Fearnside (1989).
23 Monbiot (1989).
24 Survival International (1989), Plumwood and Routley (1982).
25 Hanbury-Tennison (1989), Matthews (1990), Cunningham (1990a, 1990b), Hemming (1990).
26 Hemming (1990).
27 Sting and Dutilleux (1989), Friends of the Earth (1989c).
28 Hecht and Cockburn (1990).
29 Caufield (1985).
30 Survival International (1989).
31 Peng (1989).
32 Tyler (1990b).
33 Tyler (1990b).
34 Peng (1989).
35 Caufield (1985, p.105).
36 Lewis (1990b).
37 Westoby (1989).
38 Mendes and Gross (1989).
39 Revkin (1990a, 1990b), Hecht and Cockburn (1989), Linden (1989).
40 Colchester (1989), Friends of the Earth (1989c).
41 Cunningham (1989).
42 Caufield (1985, pp.92–3).

6 POSSIBLE SOLUTIONS

1 Such as Sting and Dutilleux (1989).
2 Prince Charles (1990).
3 Myers (1989b).
4 Earthlife Foundation (1985, p.20).
5 Eyre (1971, p.123) .
6 Prince Philip (1988 p.48).
7 Walker (1986, p.471).
8 Tyler (1990b) .
9 Plumwood and Routley (1982).
10 Barney (1980, p.154).
11 World Wide Fund for Nature (1988).
12 Goldsmith and Hildyard (1990).
13 Earthlife Foundation (1985).
14 Plumwood and Routley (1982).
15 *The Economist* (1989).
16 Di Castri *et al.* (1984).
17 Allen (1980).
18 World Wide Fund for Nature (1988).
19 World Wide Fund for Nature (1988).
20 Elsworth (1990).
21 World Wide Fund for Nature (1988).
22 White (1989).
23 Earthlife Foundation (1985).
24 Horta (1991).
25 World Resources Institute (1986).
26 Sutton *et al.* (1983).
27 World Resources Institute (1986, p.66).

28 Sutton *et al.* (1983).
29 World Wide Fund for Nature (1988).
30 World Wide Fund for Nature (1988, p.3).
31 Sutton *et al.* (1983).
32 World Resources Institute (1986).
33 World Wide Fund for Nature (1988).
34 World Wide Fund for Nature (1988).
35 Goldsmith and Hildyard (1990).
36 Gradwohl and Greenberg (1988).
37 Goldsmith and Hildyard (1990).
38 Tyler (1990b).
39 *The Economist* (1989).
40 Sullivan (1990).
41 Wright (1991).
42 World Wide Fund for Nature (1988).
43 World Wide Fund for Nature (1988).
44 Peters *et al.* (1989).
45 Hartshorn (1990).
46 Oldfield (1989).
47 Friends of the Earth (1989a).
48 Cleves (1990).
49 World Wide Fund for Nature (1988).
50 Friends of the Earth (1989a).
51 Friends of the Earth (1989c).
52 Oldfield (1989).
53 World Resources Institute (1988).
54 World Wide Fund for Nature (1988).
55 Oldfield (1989).
56 World Resources Institute (1987).
57 World Wide Fund for Nature (1988).
58 World Resources Institute (1988).
59 World Resources Institute (1986).
60 Lewis (1990b).
61 World Resources Institute (1990).
62 Meyer (1990b).
63 World Resources Institute (1988).
64 Meyer (1990a).
65 Meyer (1990b).
66 Pearce (1989).
67 *The Economist* (1988).
68 Independent Commission on International Humanitarian Issues (1986).
69 Quoted in Gupta (1988).
70 Reid (1989).
71 World Wide Fund for Nature (1988).
72 Sutton *et al.* (1983).
73 Prince Charles (1990).

REFERENCES

Aiken, S.R. and C.H. Leigh (1986) Land use conflicts and rainforest conservation in Malaysia and Australia. *Land Use Policy* 3: 161–79.

Allen, J.C. and D.F. Barnes (1985) The causes of deforestation in developing countries. *Annals, Association of American Geographers* 75: 163–84.

Allen, R. (1980) *How to Save the World*. Kogan Page, London.

Appleton, L. (1989) Trade-off. *BBC Wildlife* (August): 504–5.

Barney, G.O. (1980) *The Global 2000 report to the President of the United States*. Pergamon, New York.

Bolin, B. (1977) Changes in land biota and their importance for the carbon cycle. *Science* 196: 613–15.

Botkin, D.B. and E.A. Keller (1987) *Environmental Studies: Earth as a Living Planet*. Merrill, Columbus, Ohio.

Bowonder, B. (1987) Environmental problems in developing countries. *Progress in Physical Geography* 11: 246–59.

Branford, S. and O. Glock (1985) *The Last Frontier: Fighting over Land in the Amazon*. Zed Books, London.

Bromley, R.J. (1972) Agricultural colonisation in the upper Amazon basin. *Tijdschrift voor Economisch en Sociale Geografie* 63: 278–94.

Bromley, R. (1981) The colonisation of humid tropical areas in Ecuador. *Singapore Journal of Tropical Geography* 2: 15–25.

Brown, L.R. *et al.* (1988) *State of the World, 1988*. Norton, New York.

Bunker, S.G. (1980) Development and the destruction of human and natural environments in the Brazilian Amazon. *Environment* 22: 14–20, 34–43.

Bunyard, P. (1974) Brazil – the way to dusty death. *The Ecologist* 4: 89–93.

Bunyard, P. (1985) World climate and tropical forest destruction. *The Ecologist* 15: 125–36.

Burley, F.W. (1985) Plan to reverse destruction of tropical forests released by international task force. *Environmental Conservation* 12: 365–6.

Button, J. (1988) *A Dictionary of Green Ideas*. Routledge, London.

Caufield, C. (1982) Brazil, energy and the Amazon. *New Scientist* (28 October); 240–3.

Caufield, C. (1983) Dam the Amazon, full steam ahead. *Natural History* 7: 60–7.

Caufield, C. (1984) Indonesia's great exodus. *New Scientist* (17 May): 21–5.

Caufield, C. (1985) *In the Rainforest*. Heinemann, London.

Clarke, W.C. (1976) Maintenance of agriculture and human habitats within the tropical forest ecosystem. *Human Ecology* 4: 247–59.

Cleves, P. (1990) Tropical rainforestry. *Geography Review* 3: 2–6.

Colchester, M. (1989) The world listens to the Indios. *Geographical Magazine* (June): 16–20.
Colinvaux, P. (1987) Amazon diversity in light of the paleoecological record. *Quaternary Science Reviews* 6: 93–114.
Colinvaux, P.A. (1989) The past and future Amazon. *Scientific American* 260: 68–74.
Collinson, A.S. (1977) *Introduction to World Vegetation*. Allen & Unwin, London.
Cook, A.G., A.C. Janetos and W.T. Hinds (1990) Global effects of tropical deforestation: towards an integrated perspective. *Environmental Conservation* 17: 201–12.
Corlett, R.T. (1988) Bukit Timah: the history and significance of a small rainforest reserve. *Environmental Conservation* 15: 37–44.
Cox, C.B, I.N. Healey and P.D. Moore (1976) *Biogeography: An Ecological and Evolutionary Approach*. Blackwell, Oxford.
Cross, A. (1990) Species and habitat: the analysis and impoverishment of variety. *Geographical Magazine* (June): 42–7.
Cunningham, P. (1989) Brazil nut uncracked. *BBC Wildlife* (August): 536–7.
Cunningham, P. (1990a) The other side of the coin. *Geographical Magazine* (September): 30–4.
Cunningham, P. (1990b) Yamomami Armageddon. *BBC Wildlife* (May): 289.
Darwin, C. (1845) *The Voyage of the Beagle*. Dent, London (1959 edition).
Denevan, W.M. (1966) A cultural-ecological view of the former aboriginal settlement of the Amazon Basin. *Professional Geographer* 18: 346–51.
Denevan, W.M. (1973) Development and imminent demise of the Amazon rainforest. *Professional Geographer* 25: 130–5.
Detwiler, R.P. and C.A.S. Hall (1988) Tropical forests and the global carbon cycle. *Science* 239: 42–7.
Di Castri, F., F.W.G. Baker and M. Hadley (eds) (1984) *Ecology in Practice*. Tycooly International Publishing, Dublin.
Dickenson, R.E. (1982) Effects of tropical deforestation on climate. *Studies in Third World Societies* 14: 411–41.
Drozdiak, W. (1982) Tackling the last frontier. *Time* (18 October): 48–55.
Earthlife Foundation (1985) *Tropical Forests: The Need for Action*. Earthlife Foundation, London.
Earthlife Foundation (1986) *Paradise Lost?* Earthlife Foundation and The Observer, London.
Eckholm, D., G. Foley and G. Bernard (1984) *Fuelwood: The Energy Crisis that Won't Go Away*. Earthscan, London.
Economist (1988) The vanishing jungle. (15 October): 25–8.
Economist (1989) (25 November): 88.
Economist (1990) (4 April): 84.
Eden, M.J. (1978) Ecology and land development: the case of Amazonian rainforest. *Transactions, Institute of British Geographers* 3: 444–63.
Ehrlich, P. and A. Ehrlich (1982) *Extinction: The Causes and Consequences of the Disappearance of Species*. Random House, New York.
Ellis, W. (1988) Brazil's imperiled rainforest. *National Geographic* 174 (6): 85–100.
Elsworth, S. (1990) *A Dictionary of the Environment*. Paladin, London.
Erwin, T.L. (1988) The tropical forest canopy: the heart of biotic diversity. pp.105–9 in E.O. Wilson (ed.) *Biodiversity*. National Academy Press, Washington, DC.
Eyre, S.R. (1968) *Vegetation and Soils*. Arnold, London.
Eyre, S.R. (1971) *World Vegetation Types*. Macmillan, London.
Eyre, S.R. (1981) *The Real Wealth of Nations*. Arnold, London.

Fearnside, P.M. (1982) Deforestation in the Brazilian Amazon: how fast is it occurring? *Ambio* 7: 82–8.
Fearnside, P.M. (1983) Land-use trends in the Brazilian Amazon region as factors in accelerating deforestation. *Environmental Conservation* 10: 141–7.
Fearnside, P.M. (1986) Agricultural plans for Brazil's Carajas Program: lost opportunity for sustainable development? *World Development* 14: 385–405.
Fearnside, P.M. (1987) Deforestation and international development projects in Brazilian Amazonia. *Conservation Biology* 1: 214–21.
Fearnside, P.M. (1989) Brazil's Balbina Dam: environment versus the legacy of the Pharaohs in Amazonia. *Environmental Management* 13 (4): 401–23.
Flenley, J.R. (1979) *The Equatorial Rainforest: A Geological History*. Butterworth, London.
Foley, G. (1985) Wood fuel and conventional fuel demands in the developing world. *Ambio* 14: 253–8.
Food and Agriculture Organisation (1982) *Tropical Forest Resources*. FAO Forestry Paper No. 30. FAO, Rome.
Fox, G.E.D. (1976) Constraints on the natural regeneration of tropical moist forest. *Forest Ecology and Management* 1: 37–65.
Friends of the Earth (1988) *Financing Ecological Destruction*. FOE, London.
Friends of the Earth (1989a) *Rain forests: protect them!* Campaign leaflet. FOE, London.
Friends of the Earth (1989b) *Help the Earth Fight Back*. Campaign leaflet. FOE, London.
Friends of the Earth (1989c) *Damming the Rainforest: Indian Peoples' Summit of Altamira*. FOE, London.
Furtado, J.I. and K. Ruddle (1986) The future of tropical forests. pp.145–71 in N. Polunin (ed.) *Ecosystem Theory and Application*. Wiley, Chichester.
Gentry, A.H. (1988) Tree species richness of Upper Amazonian forests. *Proceedings of the US National Academy of Sciences* 85: 156–9.
Gentry, A.H. and J. Lopez-Parodi (1980) Deforestation and increased flooding of the Upper Amazon. *Science* 210: 1354–6.
Goldsmith, E. and N. Hildyard (1990) *The Earth Report 2*. Mitchell Beazley, London.
Golley, F.B. (ed.) (1983) *Tropical Rain Forest Ecosystems*. Elsevier, Amsterdam.
Gomez-Pompa, A., S. Vazquez-Yanes and S. Guevara (1972) The tropical rainforest: a non-renewable resource. *Science* 177: 762–5.
Goodland, R.J.A. (1980) Environmental ranking of Amazonian development projects in Brazil. *Environmental Conservation* 7: 9–26.
Goodland, R.J.A. and J. Brookman (1977) Can Amazonia survive its highways? *The Ecologist* 7: 376–80.
Goudie, A. (1984) *The Human Impact*. Blackwell, Oxford.
Goudie, A. (1989) *The Nature of the Environment*. Blackwell, Oxford.
Gradwohl, J. and R. Greenberg (1988) *Saving the Tropical Forests*. Earthscan, London.
Grainger, A. (1980) The state of the world's tropical forests. *The Ecologist* 10: 6–52.
Grainger, A. (1984) Quantifying changes in forest cover in the humid tropics. *Journal of World Forest Resource Management* 1: 3–62.
Grainger, A. (1988) Estimating areas of degraded tropical lands. *International Tree Crops Journal* 5: 31–61.
Gray, A. (1990) Indigenous people and the marketing of the rainforest. *The Ecologist* 20: 223–8.
Guppy, N. (1984) Tropical deforestation: a global view. *Foreign Affairs* 62: 928–65.

Gupta, A. (1988) *Ecology and Development in the Third World*. Routledge, London.
Hadley, M. (1989) *Rain Forest Regeneration and Management*. Pantheon, New York.
Hall, A. (1989) *Developing Amazonia: Deforestation and Social Conflict in Brazil's Carajas Programme*. Manchester University Press, Manchester.
Hammond, A.L. (1977) Brazil explores its Amazon wilderness. *Science* 196: 513–15.
Hanbury-Tenison, R. (1989) Innocents in the line of fire. *The Sunday Times Magazine* (26 February): 42–4.
Hartshorn, G. (1990) Natural forest management by the Yanesha Forestry Co-operative in Peruvian Amazonia. In *Alternatives to Deforestation*. Columbia University Press, New York.
Hecht, S. (1980) Deforestation in the Amazon Basin: magnitude, dynamics and soil resource effects. pp.61–100 in V.H. Sutlive *et al.* (eds) *Where Have All the Flowers Gone? Deforestation in the Third World*. Studies in Third World Societies 13, Williamsburg, Va.
Hecht, S. (1990) The sacred cow in the green hell. *The Ecologist* 20: 229–35.
Hecht, S. and A. Cockburn (1989) Defenders of the Amazon. *New Statesman and Society* (23 June): 16–21.
Hecht, S. and A. Cockburn (1990) *The fate of the forest: Developers, Destroyers and Defenders of the Amazon*. Penguin, London.
Hekstra, G.P. (1989) Global warming and rising sea levels: the policy implications. *The Ecologist* 19: 4–15.
Hemming, J. (1972) Robbed of a future: the Brazilian Indians. *The Ecologist* 2: 7–10.
Hemming, J. (1990) Invaded by gold-diggers. *Geographical Magazine* (May): 26–30.
Henderson-Sellers, A. and V. Gornitz (1984) Possible climatic aspects of land cover transformation. *Climatic Change* 6: 231–57.
Hiroaka, M. and S. Yamamoto (1980) Agricultural development in the Upper Amazon of Ecuador. *Geographical Review* 52: 423–45.
Holm-Nielson, L.B., H. Balslev and I. Nelson (eds) (1989) *Tropical Forests and Botanical Diversity*. Academic Press, London.
Horta, K. (1991) The last big rush for the green gold. *The Ecologist* 21: 142–7.
Hurst, P. (1989) *Rainforest Politics: The Destruction of Forests in South-East Asia*. Zed Books, London.
Independent Commission on International Humanitarian Issues (1986) *The Vanishing Forest*. Zed Books, London.
International Earthcare Centre (1983) Fact sheet on tropical rainforests. *Environmental Conservation* 10: 71–3.
Jackson, I. (1989) *Climate, Water and Agriculture in the Tropics*. Longman, London.
Jager, J. (1986) Climatic change: floating new evidence in the CO_2 debate. *Environment* 28: 6–9, 38–41.
Janzen, D.H. (1973) Tropical agroecosystems. *Science* 182: 1212–19.
Johnson, C., R. Knowles and M. Colchester (1989a) *Rainforests: Land Use Options for Amazonia*. Oxford University Press, Oxford.
Johnson, C., R. Knowles and M. Colchester (1989b) *Rainforests: Land Use Options for Amazonia – Resource Pack*. Oxford University Press, Oxford.
Johnston, R.J. and P.J. Taylor (eds) (1988) *A World in Crisis?*. Blackwell, Oxford.
Jordan, C.F. (1985) *Nutrient Cycling in Tropical Forest Ecosystems*. Wiley, Chichester.
Jordan, C.F. (1989) *An Amazonian Rain Forest*. Pantheon, New York.
Kartawinata, K. *et al.* (1981) The impact of man on a tropical forest of Indonesia. *Ambio* 10: 115–19.

Kellogg, C.E. (1963) Shifting cultivation. *Soil Science* 95: 221–30.
Lal, R. (1989) Soil degradation and conversion of tropical rainforests. pp.137–54 in R. Botkin *et al.* (eds) *Changing the Global Environment*. Academic Press, London.
Lamb, D. (1989) *Exploiting the Tropical Rain Forest*. Pantheon, New York.
Lanly, J.P. (1982) *Tropical Forest Resources*. United Nations Food and Agriculture Organisation, Rome.
Law, N. and D. Smith (1987) *Decision Making Geography*. Hutchinson, London.
Lemonick, M. (1987) The heat is on. *Time* (19 October): 21.
Lewis, D. (1990a) Fresh air in the intemperate zones. *BBC Wildlife* (June): 386–91.
Lewis, D. (1990b) You can't see the wood for the trees. *Geographical Magazine* (July): 22–7.
Linden, E. (1989) Playing with fire. *Time* (18 September): 44–50.
Longman, K.A. and J. Jenik (1987) *Tropical Rain Forest and its Environment*. Longman, London.
Lovejoy, T.E. (1989) Deforestation and the extinction of species. pp.91–8 in A. Botkin *et al.* (eds) *Changing the Global Environment*. Academic Press, London.
Mabberly, D.J. (1983) *Tropical Rain Forest Ecology*. Blackie, Glasgow.
McDonagh, S. (1986) *To Care for the Earth: A Call to a New Theology*. Chapman, London.
McIntyre, L. (1988) Last days of Eden. *National Geographic* 174 (6): 50–62.
Maguire, A. and J. Brown (1986) *Bordering on Trouble: Resources and Politics in Latin America*. Adler, New York.
Malingreau, J.P., G. Stephens and L. Fellows (1985) Remote sensing and forest fires: Kalimantan and North Borneo in 1982–83. *Ambio* 14: 314–21.
Mather, A.S. (1987) Global trends in forest resources. *Geography* 72: 1–15.
Matthews, E. (1983) Global vegetation and land use: new high-resolution data bases for climatic studies. *Journal of Climate and Applied Meteorology* 22: 474–87.
Matthews, G. (1990) A tribe dying in the gold rush. *The Times Review* (27 January): 33–5.
Melillo, J.M. *et al.* (1985) A comparison of two recent estimates of disturbance in tropical forests. *Environmental Conservation* 12: 37–40.
Mendes, C. and T. Gross (1989) *Fight for the Forest: Chico Mendes in his Own Words*. Third World Publications, New York.
Meyer, A. (1990a) The campaign continues. *Geographical Magazine* (February): 14–15.
Meyer, A. (1990b) How to tell the green from the camouflage. *BBC Wildlife* (January): 5.
Mills, S. (1990) Alone in the longhouse. *BBC Wildlife* (September): 617–20.
Monbiot, G. (1989) The transmigration fiasco. *Geographical Magazine* (May): 26–30.
Moore, P. (1986) What makes rainforests so special? *New Scientist* (2 August): 38–40.
Moore, P.D. (1990) The exploitation of forests. *Science and Christian Belief* 2: 131–40.
Myers, N. (1980a) The present status and future prospects of tropical moist forests. *Environmental Conservation* 7: 101–14.
Myers, N. (1980b) *The Sinking Ark*. Pergamon, Oxford.
Myers, N. (1981) The hamburger connection: how Central America's forests became North America's hamburgers. *Ambio* 10: 3–8.
Myers, N. (1983) The tropical forest issue. pp.1–28 in T. O'Riordan and R.K. Turner (eds) *Progress in Resource Management and Environmental Planning* 4.

Myers, N. (1984) *The Primary Source: Tropical Forests and our Future*. Norton, New York.
Myers, N. (ed.) (1985a) *The Gaia Atlas of Planet Management*. Pan, London.
Myers, N. (1985b) Tropical deforestation and species extinctions: the latest news. *Futures* 17: 451–63.
Myers, N. (1988a) Threatened biotas: 'hotspots' in tropical forests. *The Environmentalist* 8: 1–20.
Myers, N. (1988b) Tropical forests – why they matter to us. *Geography Review* 4: 16–19.
Myers, N. (1988c) Tropical deforestation and climatic change. *Environmental Conservation* 15: 293–8.
Myers, N. (1988d) Tropical forests: much more than stocks of wood. *Journal of Tropical Ecology* 4: 209–221.
Myers, N. (1989a) The future of forests. pp.22–40 in L. Friday and R. Laskey (eds) *The Fragile Environment*. Cambridge University Press, Cambridge.
Myers, N. (1989b) *Deforestation Rates in Tropical Forests and their Climatic Implications*. Friends of the Earth, London.
Myers, N. (1990) Review of *The Rainforests: A Celebration. Geographical Magazine* (January): 55.
Myers, N. and R. Tucker (1967) Deforestation in Central America: Spanish legacy and North American consumers. *Environmental Review* 11: 55–71.
National Academy of Sciences (1982) *Ecological Aspects of Development in the Humid Tropics*. NAS, Washington, DC.
Nortcliff, A. (1989) A review of soil and soil-related constraints to development in Amazonia. *Applied Geography* 9: 147–60.
North, R. (1986) *The Real Cost*. Chatto & Windus, London.
Ooi Jin Bee (ed.) (1983) *Natural Resources in Tropical Countries*. Singapore University Press, Singapore.
Oldfield, S. (1989) The tropical chainsaw massacre. *New Scientist* (23 September): 54–7.
Park, C.C. (1991) *Environmental Hazards*. Macmillan, London.
Park, C.C. (1992) *Caring for Creation*. Marshall Pickering, London.
Pearce, F. (1989) Kill or cure? Remedies for the rainforest. *New Scientist* (16 September): 40–3.
Pears, N. (1977) *Basic Biogeography*. Longman, London.
Peng, M.K.K. (1989) Developed to death. *BBC Wildlife* (December): 785.
Peters, C.M., A.H. Gentry and R.O. Mendelson (1989) Valuation of an Amazonian rainforest. *Nature* 339: 655–6.
Plumwood, V. and R. Routley (1982) World rainforest destruction – the social factors. *The Ecologist* 12: 4–22.
Poore, D. (1990) *No Timber without Trees: Sustainability in the Tropical Forest*. Earthscan, London.
Prance, G.T. (ed.) (1986) *Tropical Rainforests and the World Atmosphere*. Westview Press, Boudler, Colo.
Prance, G. (1989) Give the multinationals a break! *New Scientist* (23 September): 62.
Prance, G.T. and T.E. Lovejoy (eds) (1985) *Amazonia*. Pergamon, Oxford.
Prescott-Allen, R. and C. Prescott-Allen (1988) *Genes from the Wild: Using Wild Genetic Resources*. Earthscan, London.
Prince Charles, HRH (1990) Now or never: the Kew Lecture. *BBC Wildlife* (June): 394–402.
Prince Philip, HRH (1988) *Down to Earth*. Collins, London.
Ramade, F. (1984) *Ecology of Natural Resources*. Wiley, Chichester.

Rand, S. *et al.* (eds) (1982) *Ecology of a Tropical Forest.* Oxford University Press, Oxford.
Rapaport, R.A. (1976) Forests and man. *The Ecologist* 6: 240–7.
Reichel-Dolmatoff, G. (1977) Cosmology as ecological analysis . . . a view from the rainforest. *The Ecologist* 7: 4–11.
Reid, W.C.V. (1989) Sustainable development: lessons from success. *Environment* 31: 6–9, 29–35.
Revelle, R. (1982) Carbon dioxide and world climate. *Scientific American* 247: 35–44.
Revkin, A. (1990a) Jungle law. *The Sunday Times Magazine* (29 July): 46–50.
Revkin, A. (1990b) *The Burning Season: the Murder of Chico Mendes and the Fight for the Amazon Rain Forest.* Collins, London.
Richards, P.W. (1970) *The Life of the Jungle.* McGraw-Hill, New York.
Richards, P.W. (1973) The tropical rainforest. *Scientific American* 229: 58–68.
Richards, P.W. (1976) *The Tropical Rain Forest.* Cambridge University Press, Cambridge.
Ridgeway, R. (1987) Rainforests: the trail of destruction. *Landscape* 3: 32–5.
Rudel, T.K. (1983) Roads, speculators and colonisation in the Ecuadorian Amazon. *Human Ecology* 11: 385–403.
Russell, J.A (1942) Fordlandia and Belterra, rubber plantations of the Tapajos River, Brazil. *Economic Geography* 18: 125–45.
Salati, E. and P.B. Vose (1983) The depletion of tropical rainforests. *Ambio* 12: 67–71.
Salati, E. and P.B. Vose (1984) Amazon Basin: a system in equilibrium. *Science* 225: 129–136.
Salati, E. *et al.* (1979) Recycling of water in the Amazon Basin: an isotopic study. *Water Resources Research* 15: 1250–8..
Sanchez, P.A. *et al.* (1982) Soils of the Amazon Basin and their management for continuous crop production. *Science* 216: 821–7.
Sayer, J.A. and S. Stuart (1988) Biological diversity and tropical forests. *Environmental Conservation* 15: 22–5.
Schneider, S.H. (1989) The greenhouse effect: science and policy. *Science* 243: 771–85.
Schultes, R.E. (1980) Amazonia as a source of new economic plants. *Economic Botany* 33: 259–66.
Scott, M. (1986) South-East Asia's forests: lost for the trees. *Far East Economic Review* (10 April): 89–90.
Scott, M. (1989) The disappearing forests. *Far East Economic Review* 143: 34–5, 38, 40–1.
Secrett, C. (1986) The environmental impact of transmigration. *The Ecologist* 16: 77–88.
Sedjo, R.A. (1989) Forests: a tool to moderate global warming? *Environment* 31: 14–20.
Silcock, L. (ed.) (1989) *The Rainforests: A Celebration.* Barrie & Jenkins, London.
Simmons, I.G. (1989) *Changing the Face of the Earth.* Blackwell, Oxford.
Singh, J.S. *et al.* (1984) India's silent valley and its threatened rainforest ecosystems. *Environmental Conservation* 11: 223–33.
Sioli, H. (1985) The effects of deforestation in Amazonia. *Geographical Journal* 151: 197–203.
Smith, N.J.H. (1980) Anthrosols and human carrying capacity in Amazonia. *Annals of the Association of American Geographers* 70: 553–66.
Smith, N.J.H. (1981) Colonisation lessons from a tropical rainforest. *Science* 214: 755–61.

Smith, N.J.H. and R.E. Schultes (1990) Deforestation and shrinking crop genepools in Amazonia. *Environmental Conservation* 17: 227–34.
Sterling, T. (1973) *The Amazon*. Time-Life Books, Amsterdam.
Sternberg, H. (1988) Amazon abundant. *Geographical Magazine* 60: 30–5.
Sting and J. Dutilleux (1989) *Jungle Stories: The Fight for the Amazon*. Barrie & Jenkins, London.
Stott, P.A. (1978) Tropical rainforest in recent ecological thought. *Progress in Physical Geography* 2: 80–98.
Sullivan, F. (1990) Nailing hardwood. *BBC Wildlife* (June): 357.
Survival International (1989) *Tribal Peoples in Indonesia*. Survival International, London.
Survival International (1990) *Yanomami*. Survival International, London..
Sutton, S.L., T.C. Whitmore and A.C. Chadwick (1983) *Tropical Rain Forest – Ecology and Management*. Blackwell, Oxford.
Swinbanks, D. (1989) Japan no help to rainforests. *Nature* 338: 606.
Townsend, J. (1990) Brazil: new geography for a new President. *Geography Review* 3: 36–8.
Tucker, C.J.B., B.N. Holben and T.E. Goff (1984) Intensive forest clearance in Rondonia, Brazil, as detected by satellite remote sensing. *Remote Sensing Environment* 15: 255–61.
Tucker, R. and J. Richards (1976) *The Tropical Rainforest*. Cambridge University Press, Cambridge.
Tyler, C. (1989) Towards a warmer world. *Geographical Magazine* (March): 40–3.
Tyler, C. (1990a) Laying waste. *Geographical Magazine* (January): 26–30.
Tyler, C. (1990b) The sense of sustainability. *Geographical Magazine* (February): 8–13.
United Nations Environment Programme (1987) *The State of the World Environment*. Blackwell, Oxford.
Veevers-Carter, W. (1987) *Riches of the Rain Forest*. Oxford University Press, Oxford.
Vesiland, P.J. (1987) Brazil: moment of promise and pain. *National Geographic* (March): 348–85.
Walker, D. (1986) Tropical rain forests. *Science Progress* 70: 461–72.
Webster, K. and D. Williams (1988) *Choosing the future: rainforest*. World Wide Fund for Nature, Godalming.
Wells, G. (1989) Observing earth's environment from space. pp.148–92 in L. Friday and R. Laskey (eds) *The Fragile Environment*. Cambridge University Press, Cambridge.
Westoby, J. (1989) *An Introduction to World Forestry*. Blackwell, Oxford.
White, D. (1989) The state of the park. *BBC Wildlife* (December): 808.
White, S. (1978) Cedar and mahogany logging in eastern Peru. *Geographical Review* 68: 394–416.
Whitmore, T.C. (1990) *Tropical Rainforests*. Clarendon, Oxford.
Williams, M. (1989) Deforestation: past and present. *Progress in Physical Geography* 13: 176–208.
Wilson, W.L. and A.D. Johns (1982) Diversity and abundance of selected animal species in undisturbed forest, selectively logged forest and plantations in Kalimantan, Indonesia. *Biological Conservation* 24: 205–18.
Woodwell, G.M. *et al.* (1978) The biota and the world carbon budget. *Science* 199: 141–6.
Woodwell, G. *et al.* (1983) Global deforestation. *Science* 222: 153–5.
World Bank (1984) *World Development Report 1984*. Oxford University Press, London.

World Resources Institute (1986) *World Resources 1986*. Basic Books, New York.
World Resources Institute (1987) *World Resources 1987*. Basic Books, New York.
World Resources Institute (1988) *World Resources 1988–89*. Basic Books, New York.
World Resources Institute (1990) *World Resources 1990–91*. Basic Books, New York.
World Wide Fund for Nature (1988) *Conservation of Tropical Forests*. Special Report 1. WWF, Gland, Switzerland.
Wright, M. (1991) Sarawak story. *Geographical Magazine* (January): 47–9.
Young, A. (1974) Some aspects of tropical soils. *Geography* 59: 233–9.

INDEX

United States of America (USA) 26, 28, 66, 68, 70, 88, 89, 129